001359982 HDQ

Jackson County Library Services
Medford, OR 97501

D0604211

DATE DUE			8/01
SEP 13 01			
NOV 06 01			
NOV 27 01			
DEC 29 01			
JAN 23 03			
JUL 0 7 2003			
MAR 31 04			
3-1-07			
GAYLORD			PRINTED IN U.S.A.

ROUTER
JOINERY

ROUTER JOINERY

Gary Rogowski

The Taunton Press

JACKSON COUNTY LIBRARY SERVICES
MEDFORD, OREGON 97501

COVER PHOTO: Scott Phillips

About Your Safety

Working wood is inherently dangerous. Using hand or power tools improperly or ignoring standard safety practices can lead to permanent injury or even death. Don't try to perform operations you learn about here (or elsewhere) unless you're certain they are safe for you. If something about an operation doesn't feel right, don't do it. Look for another way. We want you to enjoy the craft, so please keep safety foremost in your mind whenever you're in the shop.

Taunton
BOOKS & VIDEOS

for fellow enthusiasts

Text © 1997 by Gary Rogowski
Illustrations © 1997 by The Taunton Press
All rights reserved.

First printing: 1997

Printed in the United States of America
10 9 8 7 6 5 4 3 2 1

A FINE WOODWORKING Book

FINE WOODWORKING® is a trademark of The Taunton Press, Inc., registered in the U.S. Patent and Trademark Office.

The Taunton Press, 63 South Main Street, PO Box 5506, Newtown, CT 06470-5506
e-mail: tp@taunton.com

Library of Congress Cataloging-in-Publication Data

Rogowski, Gary
    Router Joinery / Gary Rogowski.
      p.   cm.
    Includes index.
    ISBN 1-56158-174-7
    1. Furniture making.   2. Routers (Tools) I. Title.
TT194.R65 1997   97-6238
684.1'04 — dc21   CIP

To Jane Hester, for her patience in the early years.

To Vincent Laurence, for helping me turn the corner.

And to Charles Hayward, from whom I learned so much.

ACKNOWLEDGMENTS

I would like to acknowledge these people for the wonderful support, encouragement, and help they've given me throughout the writing of this book:
Helen Albert, Terry Anderson, Elliot Apatov, Joel Aycock, John Eric Byers, Michael DeForest, Ruth Dobsevage, Lucy Larkin, Vincent Laurence, David Minick, Bob Rashkin, Brooke Stephens, Amy Thompson, and Craig Umanoff.

Thanks to my students, for their great curiosity about the craft of woodworking.

My thanks as well to these folks for their aid and information:
Troy Bowman at Freud, Chris Carlson at Bosch, Art Carpenter, Brian Corbley at Amana, Heidi Couch and Matt Nelson at Hitachi, David Keller of Keller & Co., Bill LaHay at *Workbench* magazine,
Carter Williams at DeWalt Tool Co., Jean Miskimon at Eisner and Associates, Rich Wedler of Micro Fence Co., Marilyn Welter at Crosscut Hardwoods, Brad Williams at CMT Tools, and the 1874 House in Portland Oregon.

And for her great belief in me and support, Ketzel Levine, party of one.

CONTENTS

INTRODUCTION

There are many ways of joining pieces of wood together—it's such a remarkably resilient material that it allows all manner of construction. Wood can be fashioned into shapes, nailed, bent and laminated, bolted and screwed or held together with wire and pitch. Some of these approaches are more temporary than others. Other methods, however, have proven themselves to be strong, beautiful, and long-lasting, especially for building fine furniture. These are the joinery techniques we're about to explore. And we're about to examine how to cut these wood-to-wood joints with a remarkably simple and adaptable tool called the router.

The router is basically a high-speed motor that can hold a variety of cutting bits. In today's woodworking shop, this tool is not merely a convenience. It is an essential power tool of the furniture maker. For the professional it's an indispensable part of a kit. And for the skilled amateur, who is no less concerned with accuracy, the router is a dependable and affordable addition to the toolbox. The router's versatility and speed in profile work, shaping wood, and template routing are unmatched.

However, the router's ability to cut joints is perhaps its most useful feature. The very word "router" suggests guiding something along a path. And this is essentially what you do with one when cutting joints, either hand held ("topside") or with the router mounted bit up in a router table. These guided cuts, if indexed, can be reproduced indefinitely—or until the bit, the motor, or you wear out.

The keys to using the router for cutting joints are three. The first is understanding what a router is capable of doing, the second is knowledge of joinery and which joints to cut for a given job, and the third is directing the router to make these cuts.

Now understanding how routers work and what their strengths and limitations are is obviously important. But you won't find this information by looking through a tool catalog. In Part I of this book, I will take a look at the various kinds of routers available and what they can do. Then I'll discuss the bits, fences, guides, and jigs you can use to direct the router to cut joints. These are the tools and accessories that I have found useful in building my furniture.

Joinery for fine furniture is based on a time-honored tradition of handwrought work, using joints that have shown their resistance to the ravages of time and water, abuse and wear, in furniture that has survived over the centuries. But understanding the principles of joinery doesn't require an engineering degree, and cutting joints doesn't require a huge collection of tools. There are only three principal systems of furniture construction: carcase (e.g., cabinets), frame and panel (e.g., doors or beds), and stool and table (e.g., chairs and tables). In Part II of this book, I will discuss the role of joinery in furniture making and explore the variety of joints available to the furniture maker within each of these systems.

Being able to guide the router predictably to make joints is the third critical step. Most people recognize a dovetail joint, but is there a best way to cut it? There are so many options for cutting joints, and each has its advantages and drawbacks. In Part III of this book, I will take a look at router-cut joinery for each of the principal systems of furniture construction, along with various setups, fixtures, and jigs you can build in your own shop.

There is no one right way of producing work or joinery. One woodworker's well-engineered jig is another's construction nightmare. I have 23 years of furniture-making experience behind me, and I know that my way is not the only way of doing things. It just works for me. Woodworking joinery follows some simple rules. Their application, however, is as varied as the makers who employ them. Everyone knows that long-grain surfaces glue together best, yet many workers use end-grain joints successfully for their own type of work. Decisions about joinery often come from weighing several factors, including time and economy. Ultimately, you, the maker, have to decide how you want to build.

If your goal is to create a structure that will last 10 years, like some cabinetry, then certain joinery techniques like stapling or nailing are appropriate and desirable. If, on the other hand, your goal is to create work that will survive the vagaries of style and the passage of time, then the classic wood-to-wood joints that have proven to be strong and long-lasting are a better choice. These methods offer their own rewards. They provide a satisfaction that comes from the process of building and goes beyond the completion of the actual piece.

Finally, what I hope this book accomplishes is to give you a framework, a way to think about furniture construction. The router offers one approach to joinery that concentrates on speed, versatility, and accuracy. With it your ability to produce precise joints is enormously increased. I hope you will find within these pages the information about routers and joinery that will help make your woodworking more efficient, your furniture stronger, and your time in the shop more enjoyable. And that this book is the one that you reach for in your shop as you continue to build through the years.

PART ONE

INTRODUCTION TO ROUTER WORK

CHAPTER ONE

ROUTERS

Routers are small but powerful tools that can transform the way you cut joints and assemble your projects. Photo by Scott Phillips.

A router is capable of accomplishing many jobs in the shop. But it really can't do everything, as some tool ads would have you believe. A router is not a saw or a wood rasp. It does not stir paint or plan your trips to the lumberyard. It will not purposely destroy wood faster than you can mill it, no matter what your experience may have been in the past. What it can do is make a variety of accurate cuts in wood if you know how to direct it properly. It can transform your shop from one where projects are knocked together with nails to a shop that can perform pattern routing, shaping, and perhaps the router's most useful capability, joinery.

There are a lot of routers on the market today. Go to your local hardware center and take a look at the portable power tools displayed there. This impressive wall of plastic and metal either fills you with delight or overpowers you with indecision. Like a kid in a candy store, you want them all, and you want them all now. But the adult in you knows that you are incapable of understanding all the differences among routers and all of their nuances. And you certainly can't afford to buy all these beauties. So off you go, babbling, to the paint and putty section of the store just to cool down.

There is basic information you can arm yourself with that will help you in picking out a router. Differences exist between types and models. If you can answer some simple questions about what you will do with the tool, you can knock out most of the confusion surrounding this purchase. Then you can approach the router

display with the calm assurance of a seasoned veteran and point to the tool of your choice with a steady finger.

ROUTER TYPES

A router is made up of a high-speed motor that's held in a base. The motor shaft is fitted so you can mount bits into it. Bits come in hundreds of different shapes, allowing you to make cuts into the middle of a piece of wood or along its edges.

What sets routers apart from other power tools is speed. Routers spin at much higher speeds than most other motors. A drill spins at speeds from 250 rpm to 2,500 rpm; a router spins at 15,000 rpm to 30,000 rpm. The reason that routers spin at such high speeds is because of the nature of the

The fixed-base router, left, has a motor that separates from the base. It moves within the base to adjust the cutting depth of the bit. The plunge router, right, has a motor that is always connected to the base. It rides on two columns that allow you to plunge the motor and bit straight down into a workpiece.

cutting work that's done with them. A router needs higher speeds to produce better cutting action and smoother results with the relatively low-horsepower motor it uses.

Bits mount into a tapered sleeve called a collet. Collets are like drill chucks except that they come in only a few shank sizes and are not adjustable. The collet fits into the motor shaft (see pp. 24-26), where it's held firmly with a collet nut that threads onto the shaft.

There are two basic types of router: the fixed-base router and the plunge router (see the photo on p. 5).

Fixed-base routers

When I started woodworking, the fixed-base router was the most common type of router around. An all-metal case surrounding about 1 hp of motor, it could be had in either black or gray. It's a little different now. Fixed-base routers now come with a variety of features and shapes in their own brand-name colors, but they all share some common traits.

Unlike the drill, which is a single unit, the fixed-base router has two parts, a motor and a base (see the drawing at left). By moving the motor up or down in the base, you can adjust the depth of the bit and its cut. The motor and base also come apart, allowing you to install and remove bits with greater ease. The motor is locked into the base with a locking knob mounted on the base.

Fixed-base routers come in three sizes: $\frac{1}{4}$ in., $\frac{3}{8}$ in., and $\frac{1}{2}$ in. Routers with only $\frac{1}{4}$-in. shank capacity have a limited usefulness because there are fewer bits available in $\frac{1}{4}$-in. shanks. Larger-diameter bits that are available with these small shanks are more prone to flexing and even breaking under too large a load. These $\frac{1}{4}$-in. routers also have the smallest motor sizes. Routers with $\frac{3}{8}$-in. shanks aren't generally used by woodworkers, and the bits can be hard to find. The workhorse routers are those with $\frac{1}{2}$-in. shank capacity. These routers accept collets of all sizes, and they have motors that have a medium- to heavy-duty rating.

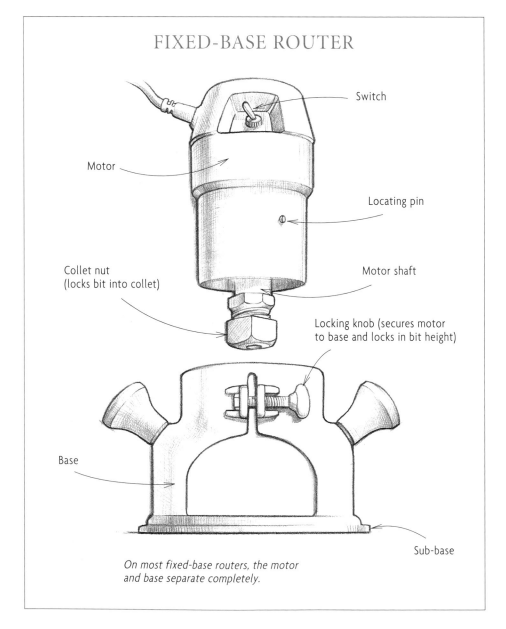

FIXED-BASE ROUTER

Switch

Motor

Locating pin

Collet nut
(locks bit into collet)

Motor shaft

Locking knob (secures motor to base and locks in bit height)

Base

Sub-base

On most fixed-base routers, the motor and base separate completely.

Height-adjustment mechanisms

There are differences in the way that fixed-base routers adjust for cutting. Several models have a small locating pin or pins mounted into the motor body. This pin either engages in a slow spiral cut made into the base or it fits right into a straight groove. The screw-action type, shown in the photo at right, lets you spin the motor like a giant screw in the base. This moves the bit up or down, exposing more or less of it for a cut. On my old Stanley router, the pin also keeps the switch in the same location so you can find it with ease.

When a fine-adjustment ring is attached to the router motor, adjustments can also be made by turning the ring. This moves the motor slowly up or down in the base for fine adjustments. When a fine-adjustment ring is attached to the base, the ring serves only as a measuring device. Each mark on the ring measures a distance of $\frac{1}{64}$ in. or smaller.

Bosch fixed-base routers use another system for fine tuning the bit height. The motor fits straight into the base until a column on the motor contacts the inclined ramp on the top edge of the base (see the photo at right). The column has an angled end on it and rides along this ramp. When you spin the motor, the column moves the motor up or down on the ramp to fine-adjust the bit.

There is a problem, however, with adjusting all fixed-base routers: The router motor loses its center in the base every time it's adjusted up or down. There is some play between the motor and the base, and this translates into slight variations

Many fixed-base routers have a small locating pin on the motor housing that lines up with a slot in the base. The slot can be a screw type (shown) or a straight groove. The pin keeps the router in line when adjusting for depth of cut.

from cut to cut. My old Stanley router suffers from this problem, and whenever I adjust the bit deeper, I am guaranteed a step in my cut because of all the play in the base.

How do you make accurate adjustments on these routers if you happen to have one of them? The key is learning the quirks of the tool and then smiling through the bad times if that fails. I try to hold my old Stanley with the same kind of pressure each time I adjust it. In this way I hope to put the same kind of deflection in it. It's a rough cure, but it works okay.

Routers that use a rack-and-pinion adjustment mechanism, such as DeWalt and the Black & Decker routers, don't have this centering problem because the rack-and-pinion mechanism moves the motor straight up and down in the base. To make this system work, the motor and base have to fit more

This fixed-base router uses a ramp on the top edge of the base and a protruding column on the motor for fine adjustments to the bit depth. As the motor is screwed in the base, the column moves against the inclined ramp and raises or lowers the motor.

tightly, so there is less side-to-side slop. While this seems like a relatively new way of adjusting a router, it's actually been around since 1957.

As shown in the photo below left, the rack is a bar with teeth on it that attaches directly to the motor. The pinion gear is located on the base. When the gear engages the rack's teeth, it translates the rotational motion of the knob on the base into up-and-down movement in the motor. The motor can be moved in tiny increments and holds its center in the base fairly precisely. The locking mechanism locks right over the rack-and-pinion gears, securing the base.

Plunge routers

Plunge routers have been around for decades, but they're finally being designed to be something other than a work hog in a large production shop. They're still made for power, but now they're being built for accuracy and precision too.

The unique design of this tool (see the drawing on the facing page) is what makes it so valuable for topside work. Plunge routers have a motor and base that do not separate, either for changing the bit or for setting the bit depth. However, the motor is spring loaded onto two columns that are attached to the base. It travels vertically along these columns as you push down on the handles of the tool.

What this means is that you can set a plunge router over a board and plunge straight down into the wood. If the router is held securely or is indexed properly with a fence

or guide, there will be none of the bit-centering problems that fixed-base routers have. Therefore the plunge router is ideally suited for mortising and other topside operations. And changing a bit on a plunge router is a simple task because there's easy access to the motor shaft.

Another advantage of the plunge router over the fixed-base router is how its depth of cut is set (see the photos on p. 10). The bit can be exposed for a full depth of cut just like a fixed-base router. The plunge router has about 2 in. to $2\frac{1}{2}$ in. of travel, and a bit can be set anywhere along the plunge stroke with the locking handle that's on the base. If you loosen that lock, the motor moves back up to its original position.

But the plunge router can also be indexed to plunge down to a predetermined depth. Instead of having to shut off the motor and change bit depths after each pass, you first set the bit to cut at full depth, then raise it up to make the first pass. Then you make a series of passes. After each pass, the bit is pushed a little deeper into the wood until the full depth of cut is achieved.

On a plunge router, cutting depth is set with an adjustable depth rod that is mounted to the motor. A turreted stop with several different height settings usually provides the end point for the travel of this bar. Fine adjustments to this setting are made in a variety of ways, depending on the router. Most have some kind of screw adjustment either to the turreted stops or to the depth rod itself.

This fixed-base router uses a rack-and-pinion depth adjustment. The rack is a column with teeth in it that is attached to the motor. The pinion gear, which is attached to the knob on the base, engages the column to move the motor up and down inside the base.

PLUGE ROUTER

Spring-loaded
plunge column

Motor

Locking handle

Lock nuts

Height rod

A plunge-router motor rides on spring-loaded columns and does not separate from the base. It can be plunged down to any depth and locked in place with the locking handle. There are also turreted stops that will limit the depth of cut.

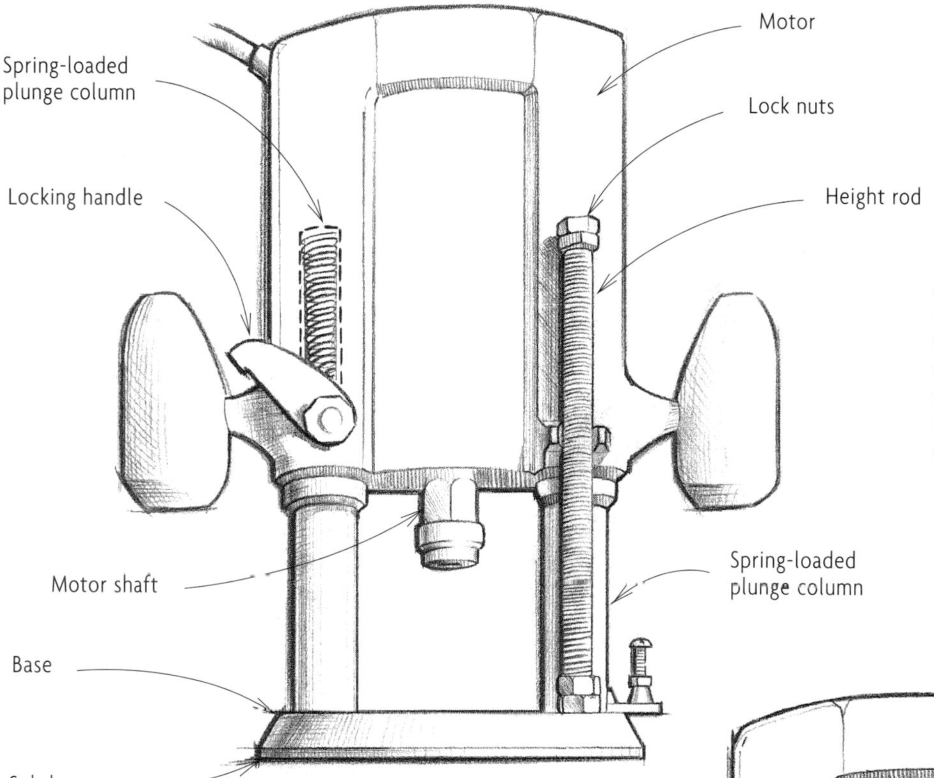

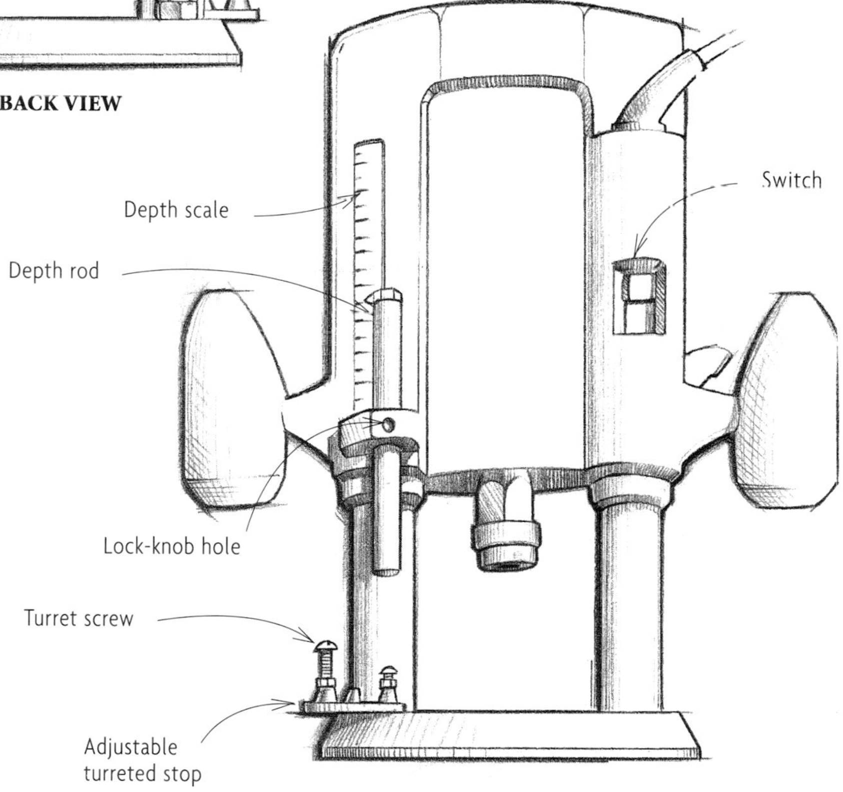

Motor shaft

Base

Spring-loaded
plunge column

Sub-base

BACK VIEW

Depth scale

Switch

Depth rod

Lock-knob hole

Turret screw

Adjustable
turreted stop

FRONT VIEW

Like a fixed-base router, a plunge router can be set up to cut by pushing down on the motor and exposing as much of the bit as is needed. The locking handle or release lever locks the motor in place.

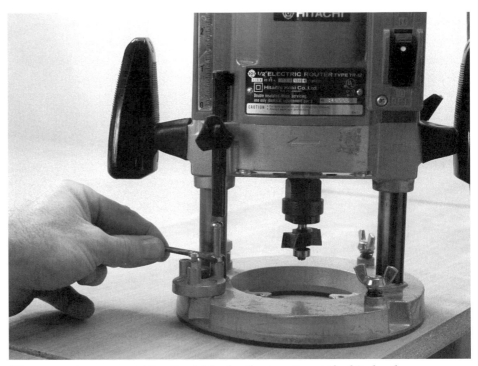

Plunge routers come with adjustable depth stops to set the bit depth. The adjustable depth rod is mounted to the motor, and the turreted stop is on the base. Fine adjustments can be made to the depth setting on the depth-stop screws.

CHOOSING A ROUTER

Classic hand tools, such as hammers and planes, never really become obsolete. When a tool is perfectly designed, there's no reason to change it. Unfortunately, this is not the case with portable power tools. New features and improvements are being introduced to them every year. You can find one that will do the job for you today, but what about the jobs you need done three years from now? How do you know which tool will continue to serve you as your abilities grow?

Which router should you buy? There's no simple answer. No single tool will do all your jobs perfectly each time. You generally have to compromise, weighing factors like versatility, price, and features. Most woodworkers buy a tool for a lifetime of use and then accept its limitations until it conks out (or they do).

My approach to buying a router is a three-step process. Think first about the kinds of tasks you want this tool to accomplish for you. Then narrow down your choices by motor features and price. Last—and most important—go out and handle the tool.

Versatility and function

What matters most if you're going to buy just one router is how many jobs it will do and how well it will do them. Being able to adapt your router for a variety of situations is critical when you're trying to come up with solutions to problems in the shop.

How do you measure versatility? One way is to consider the main types of work a router can perform: edge profiling, template routing, inlay work, and joinery. Almost any router is capable of profiling an edge detail. You just have to set the depth of cut on a bearing-mounted bit and then move the router in the right direction. But if you do a lot of profiling and shaping work, bigger is definitely not better. A few hours spent profiling edges will give you and your arms a deep appreciation for the beauty of a lightweight design. A 3-hp plunge router is versatile enough to do the work, but using it for edge work is overkill. It's not the right router for the job. For profiling, a lighter-weight fixed-base router has enough power to do the job. Template routing and inlay work also require a deft hand, so I prefer a fixed-base router for this sort of work too.

Is a fixed-base router the best choice for joinery? That depends on the joints being cut. If I were cutting a lot of joints on a router table, I would use a fixed-base router in it and buy an extra router base to do the occasional topside job. The alternative of buying a plunge router and putting it in a router table has no appeal to me. It's not what the tool was designed to do. Plunge routers are heavy, and the weight will eventually cause the router table or insert plate to sag. And with a large baseplate mounted on a plunge router for both router-table and topside use, you would have even more weight to lug around whenever you wanted to work topside or change a bit.

If you have a lot of mortises to cut, then a plunge router is clearly the better choice. It far surpasses the fixed-base router for accuracy when cutting mortises throughout the depth of a cut. It's simpler to set the depth and move through a series of passes too. And when you pull out that big profiling bit, the soft-start, electronic motors in a plunge router (see the discussion on pp. 12-13) also have an advantage over the smaller-horsepower motors in a fixed-base router.

Ideally every shop should have two routers. The fixed base router would be used in the router table. With the addition of an extra base, it could also handle profiling, inlay, and edge work where weight and handling are considerations. The plunge router could do what it does best, which is plunge down into wood or handle those situations where a large-diameter bit needs a slower-starting motor.

Motor features

Motor-related features to consider when shopping for a router include speed, horsepower, and amperage. Also decide whether soft-start and variable-feed capabilities would be useful to you. Don't concern yourself with motor size if a router has the other features you're looking for. Most routers in the same class (light duty or heavy duty) are close enough in power that there's little difference between them.

Size, speed, and power

Motor size is probably the first thing many woodworkers think of when considering a power-tool purchase. But let's get this nonsense out of the way first.

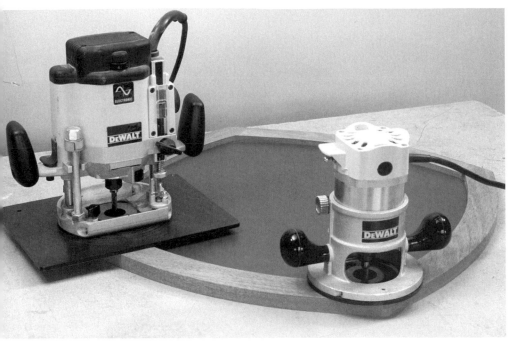

Which tool would you pick up first to make a critical cut or do precise work? A plunge router mounted with a router-table baseplate, left, has a limited usefulness compared to a fixed-base router, right.

Motor ratings are as misleading as sports statistics. One router may come touted with 30,000 rpm speed, making you think that high speed means greater power. Not so. If you look at a catalog of routers, you'll notice that the smallest routers spin the fastest, while the big hogs turn at considerably slower speeds. This is because the underpowered routers need to turn faster to get through the material. A typical laminate trimmer needs 30,000 rpm to do its job. Heavy-duty routers have the oomph to plow through wood at speeds of 20,000 rpm because of their greater power and torque. So be careful with the numbers on these tools. Higher speeds don't mean greater power.

Horsepower ratings are similarly misleading. Horsepower is a measure of a motor's output, but these ratings are done at no-load speeds under optimal conditions, not in the middle of a cut. When a router is put under a load, its speed drops immediately. You can hear a motor bog down under too heavy a pass. Amperage ratings give you a much better sense of the power of the motor. Very simply put, the higher the amps, the more power the tool will have. When looking at otherwise comparable tools, go for the one with higher amps. Greater power will come along with it.

Soft-start/variable-speed motors

As routers have replaced shapers in the small shop, new bits have been developed that do the kind of work shapers used to do. These large-profile bits have huge diameters and put a big strain on a little motor, especially when the motor first kicks on. The outer edge of the bit has to travel much faster to keep up with its center. Soft-start motors eliminate that kick to the bearings. They also reduce the impact on your hands and wrists when a powerful motor is turned on. Unfortunately soft-start motors show up only on plunge routers; there's still plenty of kick to some fixed-base motors.

A variable-speed adjustment on the motor allows you to use large profiling bits with less fear of an accident (please notice I said less fear, not all fear). The electronically speed-controlled motors now being used on plunge routers also give the motor constant power under these big loads.

Do you really need these features? I'm not so sure. If fine-tuning my motor speed for every known species of wood and kind of cut could really produce smoother results, then I should be sanding less and saving all sorts of time, the router instruction manuals hint. Sadly, this is not the case. By some miracle, most woodworkers have been getting by for years with just turning on their routers and going to work, without worrying about which speed their bog oak needs to be cut at. Soft-start motors and electronic variable-speed adjustment are standard features now on plunge routers, so I guess they are welcome. But they are far from critical to the overall performance of a router.

Handling

Design-related features such as depth-adjustment mechanisms, motor and plunge locks, and handle style are best evaluated when you use the tool. You should not buy a tool untouched. Before buying any router, pick it up, feel its heft. Is it easy to hold onto, or does the weight of the tool throw you off? Check out the handles. How do those fit your hands? It doesn't matter if the manufacturer thinks you should get one kind of handle if it doesn't feel right to you.

A woodworker will get used to one tool and its design, and everything else will feel just a little wrong. Some workers hold onto a router base with one hand jammed between a handle and the base. Others who have and use their opposable thumbs will look for handles that allow them to grip the

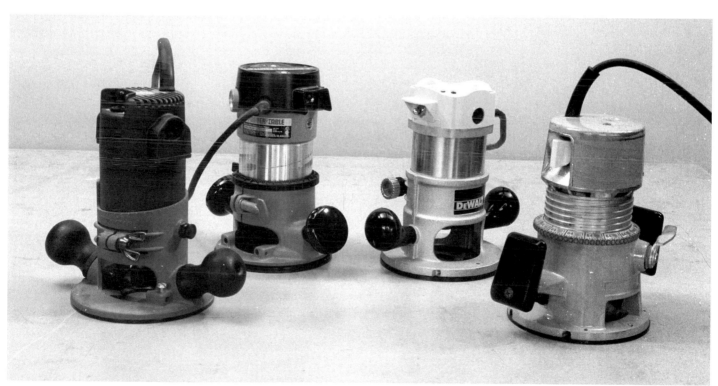

All these fixed-base routers can cut wood, but which one fits your hands best? The only way to know is to pick them up and use them. They all have differences in handle shape, motor adjustments, and switch and power-cord placement. These things matter when the tool is in your hands in the shop.

router in a more natural—and safer—fashion. Sometimes you won't have two hands to hold onto a router with, and often you will reach to pick up a tool with just one hand. Check to see if you can do this with each router you consider. A router handle should be shaped so that you're able to hold onto a heavy tool with only a little effort and with no fear of having the tool slip from your grip.

You also have to be able to use this tool with your hands. You pick it up, move it around, place it on precarious edges to do both coarse and meticulous work. So turn the router on, listen to it run, and try it out on a test cut. Don't let someone else show you how easy it is to use

and adjust. Forget about not knowing how to hold it perfectly and grab onto it and use it.

How do you install and remove bits? Are the collets substantial? Can you adjust the bit depth easily on a fixed-base router and lock the motor and base together without any fuss? If you really have to screw a lock down hard when you first buy a tool, the lock will require even more cranking to hold as it gets older. The day will come when you're using pliers to grab onto the handle.

A plunge router should have a smooth plunging action with a very positive lock. Bit-depth settings and fine adjustments should be simple and accurate. Comparing the fine tuning of depth settings can often help you decide between one model and another.

Woodworkers are notoriously frugal, but remember that when you buy a tool that you'll have it for a good long while. Don't skimp on a router just to save a few dollars. If you amortize that savings out over the usual 10-year lifetime of a tool, it can amount to only a few dollars per year.

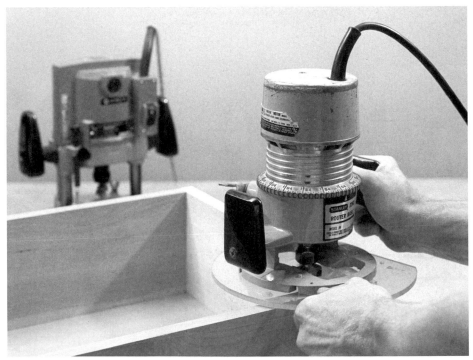

Keeping a firm grip on a power tool is essential to your safety and to the performance of the tool. How well you can hang onto a router will determine the range of jobs it will be suitable for. For example, a cut on a thin edge is easier with a fixed-base router, as shown here, than with a heavy plunge router.

BITS AND COLLETS

Router bits are held in place by the collet. They are easily inserted and removed. Photo by Jim Chiavelli.

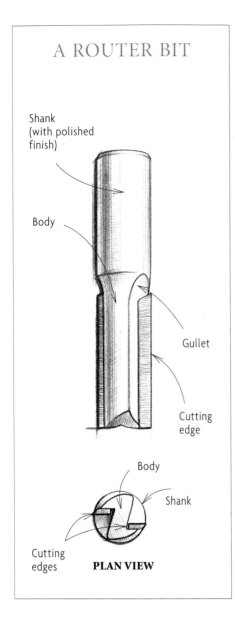

A ROUTER BIT

Shank
(with polished
finish)

Body

Gullet

Cutting
edge

Body

Shank

Cutting
edges

PLAN VIEW

Once you have a router, you'll need bits for it. Without them your router is useless. And without a collet holding the bits snug and true, your work can be inaccurate or, at worst, dangerous.

The wide variety of cuts you can make with router bits gives the tool its versatility. But there are hundreds of bits to choose from, designed to cut anything from wood to fiberglass. What purpose can all these bits have? Which ones do the best job? Rest assured that these bits are being offered because someone out there is buying them. Differences among bits can also be judged, if not measured. The issue of which to buy depends on the work you want to do and on how much you are willing to spend.

ROUTER-BIT DESIGN

Whatever their size and configuration, all bits share some design similarities. As shown in the drawing at left, all router bits consist of a body, a shank, and one or more cutting edges. The body and shank are usually made of one piece of tool steel or carbide steel. The cutting edge can be made of high-speed steel, brazed-on carbide tips, or solid carbide. The body has gullets cut in it for chip removal.

High-speed-steel bits are the least expensive. They can be ground to hold a very sharp edge, but the steel is soft. The edge breaks down fast, especially in abrasive materials such as composite board or plywood. High-speed-steel bits should be sharpened after each use, and care should be taken not to draw their temper by overheating them in a cut.

Carbide-tipped bits are the bits most commonly used and easiest to buy through catalogs and stores. Carbide is an alloy of carbon and metal powders that are fused together, yielding a material that is hard and resilient but not too brittle. The edge on a carbide-tipped bit, which is brazed onto a tool-steel shank and body (for more on the manufacturing process, see the sidebar on the facing page) will last more than 20 times as long as a high-speed-steel edge. (The finer the metal powders, the longer the edge will last.) Larger particles, only microns bigger than those used in the best bits, will break off more easily than small ones. In a good mix of carbide, as long as those particles stay bound together, the cutter will resist heat and holds its edge. As wear, excessive pounding, and heat buildup start to take place, small bits of the carbide flake off, and the bit gets dull. When this happens, take the bit to a professional for sharpening. If large chips break off (see the photo on the facing page), the bit may be beyond repair.

Better-quality carbide-tipped bits use more carbide so they're more expensive. With each sharpening, part of the face of the cutting edge is removed. The more carbide you start with, the more sharpenings can be made and the longer your bit will last.

Solid-carbide bits are the most expensive bits of all. It takes more high-priced carbide to make them and more time to grind them to a specific shape with diamond cutters. Solid carbide is usually reserved for special bits, such as spiral-flute bits or flush-trimming

HOW CARBIDE-TIPPED BITS ARE MADE

First, a blank of heat-resistant tool steel is turned slightly oversized for a shank and body. Then gullets are cut into the body of the bit. These chip pockets provide room for the cutting edge at a specific angle. The hook angle, as it's called, is made at a precise setting for each type of bit. The body of the bit behind the tips is also relieved so there's less friction as it comes through the cut.

If the bit is small enough for heat buildup to be a problem, the bit will be heat-treated next to give it some toughness. (A large-bodied bit doesn't get heat-treated, so the steel retains its resiliency.) The bit is then washed to remove any contaminants, and the carbide tips are brazed on. After sandblasting to remove debris, the bit gets a coating of a rust preventive or a Teflon-like substance to make it easier to remove built-up resins and burnt wood.

The shank and tips are ground next. Shank grinding is one of the most critical parts of the manufacturing process. The carbide tips could be perfectly ground, but if the bit doesn't spin concentrically, all that work is wasted. A ½-in. bit will have its shank ground to within 0.0005 in. under a full ½ in. Next, the edge profile and diameter are ground to specifications. If a bearing goes onto the bit, then the bit end is drilled and tapped for that as well. All this work is then inspected and checked for quality.

Most router bits are designed for general use, but they don't have to be. Theoretically, you could call up your bit maker and say I want a bit to do this and this, make me one. And if he wanted to, he could do this. He'd ask you what you're cutting, what your feed rate and motor speed will be, who will be operating the tool late in the afternoon, and how much money you have.

And then he could give you the specs and cutting angles you need and blend his carbide to match that. Then, with a dedicated application, he could engineer a bit for extended life and optimal cutting features. It would probably cost you your house, but it would be a sweetie. Unless you dropped it on the floor, of course, and chipped the carbide.

---

bits. They can hold an edge up to six times longer than carbide-tipped bits. Solid-carbide bits are also stiffer so bit deflection is minimal, but because they're more brittle they're also more likely to break under very heavy loads.

Bits are available in different shank diameters: ¼ in., ⅜ in., and ½ in. If you have a choice within the same profile, the ½-in. shank is the best option. Under the strain of a cut, ¼-in. shanks have a greater tendency to flex and chatter. They also can bend or break if not seated properly in a collet or if they're stressed by too large a cut. With ½-in.-shank bits, breakage is rarely a problem. And ⅜-in. shanks are as rare as ⅜-in. routers.

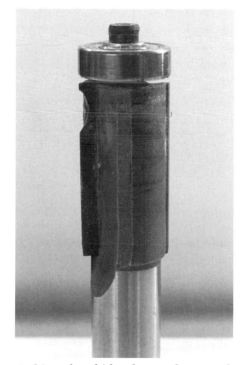

A chipped carbide edge can be caused by operator error or by flaws in the carbide—perhaps bad brazing or a poor carbide mix. Or maybe the feed rate was a little too fast.

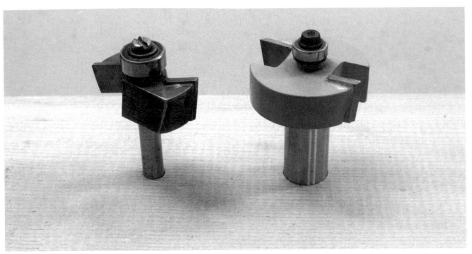

At left, a standard ⅜-in. rabbeting bit with a ¼-in. shank; at right, an anti-kickback rabbeting bit of almost the same diameter with a ½-in. shank. Anti-kickback bits limit the amount of wood that can be cut, effectively slowing you down so you can't overfeed the wood and have the bit spit back at you.

Another bit feature is the anti-kickback design found on larger profile bits (see the photo above). These have been made for European markets for a while, but are relatively new to the American market. This type of safety bit has more body left behind the cutting edge. This prevents feeding too quickly into a piece with the potential of a bit kicking back on you. While the safety design is admirable, a careful and controlled cut will be safe without this feature.

Bit quality

You can't tell which grade of carbide a manufacturer has used by looking at the bit, unless you happen to carry around an electron microscope. The grade of carbide used in a bit cannot be seen, so you have to rely on the good name of the manufacturer or catalog. Better yet, talk to others who have had experience with a brand.

The better manufacturers of carbide-tipped bits use comparable grades of carbide, good brazing techniques, and superior machining. Nevertheless, you can learn to distinguish a good bit from a cheap one. Look first at the quality of the brazing. Check for voids or holes in the braze or splattering around it. If you find any, the manufacturer may have skimped on quality control. The bit may be fine. But if the carbide tip is not firmly attached to the body, that could mean chunks of carbide flying about your shop.

Also check the quality of the grind on both the cutting edge and the bit shank. This is important for one simple reason. No manufacturer is going to waste time putting a beautifully smooth grind on lousy carbide. Good carbide, however, is worth the effort that a smooth grind takes. A pencil run along the edge of a carbide tip should move effortlessly along. Really visible grind marks appear to the naked eye as tiny ridges and look like the craters of the moon under close inspection. A poorly ground edge will give poor results. A coarse grind will heat more quickly and dull faster. Bits can be made more cheaply if the manufacturer doesn't work to high tolerances, but the bits won't stay as sharp.

BIT TYPES

In the whole grand world of bits there are two broad categories: unguided and guided (see the photo on the facing page). Unguided bits are just a shank and a cutting edge. Guided bits have either a bearing or pilot shaft on their shank that directs their cut.

Unguided bits

Unguided bits can be used for straight or profile cuts, edge work, dovetailing, or grooving, but the router has to be directed by a template or a fence for these bits to be accurate and safe. Cuts can be made anywhere on a board—either along an edge or into the middle of the wood.

Straight bits

The most common unguided bit is a straight bit. These come in specific lengths, and in cutting diameters from 1/8 in. up to 1 3/4 in. Straight bits usually have two cutting edges, or flutes. Some come with center-cutting ability (with a cutting edge ground in on the bottom) so they can be plunged more easily into stock. The body of the bit is relieved behind the flutes.

Straight bits with two flutes give the smoothest results because of the balance that two cutting edges provide. For production work that requires a high feed rate and fast chip removal, single-flute bits are often used. Occasionally three- and four-flute slotting cutters turn up, but only because these kinds of bits have large cutting diameters. Otherwise, with so many flutes, the bit would end up cutting its own chips, turning them into dust and dulling its edges.

Spiral-flute bits

The only drawback to a straight bit is that the cutting edges act like broad, flat chisels chopping into the wood through a whole line of connected fibers at once (see the drawing on p. 20). Since only one cutting edge at a time may be in contact with the work, the bit may vibrate and chatter. Taking a cue

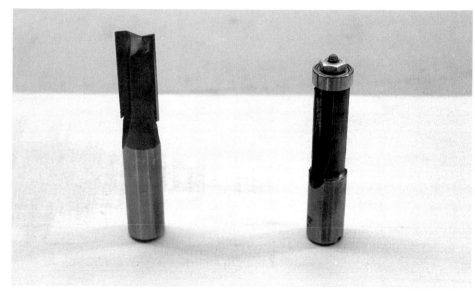

Router bits may be unguided or guided. The straight bit on the left is unguided and can make cuts anywhere on a board. The flush-trimming bit on the right uses a bearing to direct its cut, which must be along the edge of a board.

from the metalworking trade, bit manufacturers began making spiral-flute bits. Because of their steep shearing angle, spiral-flute bits slice the wood for a cleaner cut. A shearing pass will also have both its cutting edges in contact with the work for most of the bit revolution. This means the bit will have less of a chance to wobble or chatter in the cut, yielding smoother results. This is exactly what a metalworking end mill does, and many woodworkers use high-speed-steel end mills as an alternative to router bits.

Like straight bits, spiral-flute bits are manufactured in high-speed-steel, carbide-tipped and solid-carbide versions. There are two types: up spirals and down spirals (see the drawing on p. 21). Up spirals have a flute designed to eject chips just as a twist-drill bit does. This makes for quicker cutting with less heat and chip buildup. However, with veneered or

STRAIGHT CUTS VS. SHEARING CUTS

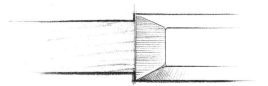

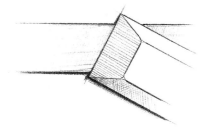

A straight cut is made straight across the wood.

A shearing cut slices through the fibers at an angle.

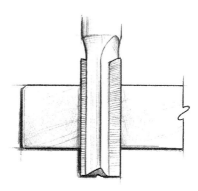

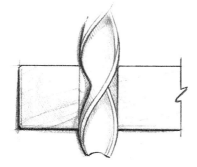

A straight bit contacts the wood in a straight line. This can sometimes lead to chatter, because for a fraction of a second, no part of the cutting edge is touching the wood.

With a spiral-flute bit, some part of the cutting edge is always in contact with the wood. As a result, the cuts will be cleaner, with less chatter.

laminate materials there can be tearout at the top of the cut. A down-spiral bit eliminates tearout by pushing the wood back into itself as the cut is made. It will also try to pull itself out of a collet if not sufficiently tightened, so install and use this type of bit with care.

Dovetail bits

Dovetail bits are another type of unguided bit. (There are bearing-mounted dovetail bits used for certain jigs, but for a variety of cuts, unguided bits are the ones you'll use most often.) Dovetail bits can be bought in several different angles, depending on the angle you want or the jig you're working with.

Guided bits

Guided bits have a pilot shaft or a bearing on them that directs or limits their cut. They can work only on the edges of boards. You can't run the bit in the middle of a board. It's that simple. Any attempt would make for some interesting smoke, but you wouldn't get far into the cut.

A pilot shaft comes only on high-speed-steel bits. This turned column of steel sticks out from the end of a bit to guide the cut. Because the pilot shaft spins just as fast as the bit, it has the annoying habit of creating friction, burning the wood in the cut.

That's why bearing-mounted bits were invented. These have a double sealed ball bearing mounted above or below the cutting edge of the bit. Bearings above the cutting edge are held in place with a retaining clip; bearings

All these bits can be used to make a ½-in. cut. In the front row, from left to right: a high-speed-steel end mill, a ¼-in. shank double-fluted straight bit, and a ½-in. shank double-fluted straight bit. In the back row, from left to right: a down-spiral carbide-tipped bit, an up-spiral carbide-tipped bit, and an up-spiral solid-carbide bit.

below the cutting edge are mounted on a pilot shaft. The shaft is cut with threads and either a setscrew or a nut holds the bearing onto it. The bearing spins as fast as the bit until it comes into contact with the wood. Then it slows down and moves as fast as your feed rate. Guide bearings are found on carbide-tipped bits and are replaceable if they wear out or if a size change is required.

Unfortunately, a worn-out bearing usually looks pretty much like a brand-new bearing. So how do you tell if a bearing is worn? The difference is like the difference between skating on ice and walking on gravel. It sounds about the same too. One is smooth and quiet; the other is slow-going and crunchy. When a bearing's lubricant gives out, the balls captured inside of it start rubbing against each other, build up heat, and finally seize together. The bearing eventually stops turning independently and spins as fast as the bit, burning the wood. If you use any type of solvent to clean your bearings, be careful that you don't inadvertently remove the lubrication. When a bearing starts to crunch, replace it.

Two types of guided bits are flush-trimming bits and rabbeting bits (see the drawing on p. 22).

(see the drawing on p. 22).

UP SPIRALS VS. DOWN SPIRALS

Bit rotation

Chip direction

An up-spiral bit pulls chips out of the hole as it moves down into a cut.

Bit rotation

Chip direction

A down-spiral bit pushes the wood fibers back into the hole as it moves down into a cut.

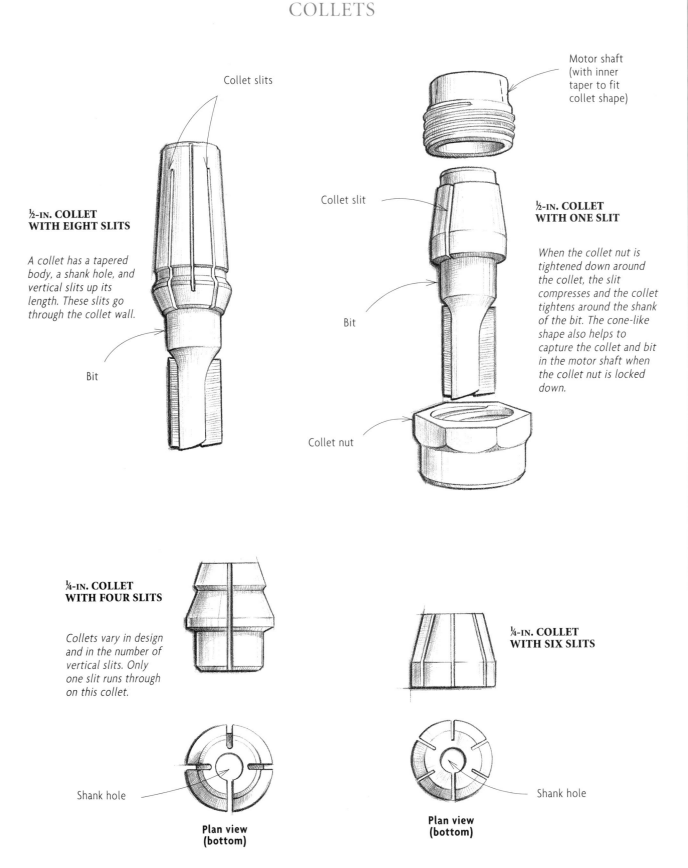

Collet slits

½-IN. COLLET WITH EIGHT SLITS

A collet has a tapered body, a shank hole, and vertical slits up its length. These slits go through the collet wall.

Bit

Motor shaft (with inner taper to fit collet shape)

Collet slit

Bit

½-IN. COLLET WITH ONE SLIT

When the collet nut is tightened down around the collet, the slit compresses and the collet tightens around the shank of the bit. The cone-like shape also helps to capture the collet and bit in the motor shaft when the collet nut is locked down.

Collet nut

¼-IN. COLLET WITH FOUR SLITS

Collets vary in design and in the number of vertical slits. Only one slit runs through on this collet.

¼-IN. COLLET WITH SIX SLITS

Shank hole

Plan view (bottom)

Shank hole

Plan view (bottom)

COLLETS

Collets in a router serve the same function as a chuck in a drill. They are not adjustable, however, and they come in only three shank sizes. Their job is to hold onto the bit without fail, and their performance is critical to the safe and accurate operation of the router.

Every router uses some type of tapered collet. Its basic design is simple (see the drawing on the facing page). Collets have a tapered body or shoulder that fits into the end of a tapered motor shaft. This cone-shaped piece of hardened steel is bored out just a little oversized so a bit shank can slide right in and out of it. It's cut with a vertical slit in one or more locations around its perimeter. This slit sometimes goes all the way through the collet. When the collet nut is tightened down, the slit or slits compress, tightening the collet around the bit shank.

Collets are manufactured to fit particular routers. Their length, the number of slits in them, their wall thickness, and other specifications vary from one make to another (see the photo above right). Only short collets used to be available, but obviously the more collet you have, the better its chances of holding onto a bit. Similarly, the more slits in a collet, the more uniform the compression on the bit and the better the holding power.

Collet sleeves

Straight sleeves are sometimes used with a collet to accommodate smaller-sized shanks. For example, you can use a 1/4-in. or 3/8-in. shank in a 1/2-in. router by fitting it into a collet sleeve, which then fits within the collet. The collet is tightened

Various collet designs in both 1/4-in. and 1/2-in. sizes. Notice the differences in length and number of slits. The collets on the left are from older fixed-base routers. The center collet is from a plunge router. The collets on the right are self-loosening collets from newer fixed-base and plunge routers.

in the usual manner. I've had no trouble with these sleeves over the years. However, more care is taken in the manufacture of a collet than a sleeve. Try to use collets rather than sleeves whenever possible for more secure holding power.

Self-loosening collets

A collet sometimes develops the nasty habit of staying locked onto a bit and stuck in the motor shaft even after the collet nut has been removed. Its steel body doesn't relax when the nut is loosened. This leads to all sorts of frustrating efforts to remedy the situation and often causes damaged bits, bearings, and motor shafts.

Self-loosening collets were designed to eliminate this problem. A small lip is placed on the inside of the collet nut. The collet either has a lip or retaining clip around its outer edge. The collet is snapped into the collet nut with one lip fitting over the other, and the entire unit is then placed into the router.

When the nut is tight, these lips are not in contact with one another. After the nut is loosened though, you'll find you get a turn or two of free play and then the nut seems to get tight again. This is because the lips are now being drawn up against each other. Another turn with the wrenches pulls the collet up with the nut as you unscrew it. This frees it from the motor shaft, loosens the collet's hold on the bit, and pulls the collet free.

The drawback to self-loosening collets, of course, is that people don't always use them properly. They require the collet and nut to be put together first and then the bit inserted. What can happen is that the collet is put into the router and a bit inserted into it, expanding its outer diameter. When the nut is placed and tightened, it can't fit down over the collet lip. When you try to loosen things up, the bit and collet stay locked in place.

Self-loosening collets are not a standard feature on all routers. Most, but not all, plunge routers use them, and only a few fixed-base routers do so. They are worthwhile, but not a critical factor if you are trying to decide on one router over another. And that second loosening that's required may sometimes seem like a nuisance. But in the long run, it's far less a bother than banging on the collet trying to loosen up a stuck bit.

Cleaning and maintenance
Collets can wear out just like any other part in a tool, and keeping them clean will increase their life expectancy. The collet shouldn't

have any dirt or debris built up on its inside or outside. The motor shaft where the collet fits should also be clean. But keep the maintenance simple: Using your finger, wipe out the hole where the collet goes before you put in a bit. It'll be fine if you keep it just that clean. Should you inherit a router with a dirt-encrusted motor shaft, clean it with a solvent-dipped toothbrush or rag. Do not sand it clean with sandpaper because you run the risk of changing its shape.

Inspect your collets regularly for scoring. The presence of circular marks on the inside of a collet means that a bit has been spinning around inside of it. Don't panic if you see a tiny score mark—it may have been caused by your spinning the bit as you seated it. But if you see some grooves in the collet or on any of your bit shafts, toss the offending collet and buy a new one. There is no point in hanging onto a collet that won't hang onto your bits. Also, never tighten down on a collet without a bit inside of it. You might distort the collet and affect its holding power.

INSERTING AND REMOVING BITS
You insert a bit by sticking its shank into the collet and tightening the collet nut; you remove a bit by loosening the collet nut and pulling the bit out. That much is pretty obvious. But there are a few more things to know.

Most woodworkers have pulled a bit just a little farther out of a collet to get that extra smidgen of length. Dangerous? But of course. It's unlikely studies have been done on this type of stupidity, but when you're in a hurry or lazy you sometimes do foolish things. What can you safely get away with? I have no desire to tempt fate, so if the bit isn't long enough, I buy one that is.

Try to have at least three-quarters of the shank of the bit in the collet. This will give the collet enough grip on the bit to prevent it from being tossed out of the router like a missile. The greater the diameter of the bit, the more shank you want inside the collet. (If you have a choice, also try to use the shorter of two bits of the same diameter. This will cut down on bit deflection.)

As to bottoming out the bit inside the motor shaft, opinions vary. I have done this for years with no ill effects on my work. But manufacturers of both bits and routers now say that you should pull a bit back up and out of a shaft about ⅛ in. because motor vibration will cause the bit to loosen. I don't buy this explanation because the motor and bearings are already transferring through the collet what little vibration they have. It's somewhat easier to believe that the bit could cock sideways as its end gets pushed down against the motor shaft, which could cause runout at its tip and vibration throughout a cut. That might cause the bit to loosen. But considering the tolerances between bit, collet, and shaft, this too seems a stretch.

Don't go by any hard-and-fast rules. If you always back a bit off ⅛ in., the collet may not be grabbing all shaft. If you find your bits loosening, pull them out of the collet a little. Should the problem continue, check your collet first and then your bit shanks. Also make sure that your motor is not moving within a fixed-base router.

After you install a bit, always check that it fits through the router-base hole and spins freely. Skip this step, which takes just a moment, and switching on the router could become a bitter learning experience. A bit that's too large will either shatter or embed itself in the edge of your router base, burning up your bearings or motor.

Tightening and loosening the collet nut

The collet nut does have to be tightened on the collet to hold in the bit. This task is important enough to do with some concentration. I tighten my collet nut down with a fair grunt and

When loosening a collet nut, hold the wrenches on the nut and shaft so they're close together. This way you can span them both with your hands and prevent any knuckle busting when the bit suddenly comes free.

A stuck bit can be loosened with a sharp rap from a thin but heavy tool. Always support the shaft well on a solid surface, and hit only the flat of the nut—never the collet or threads of the shaft.

good amount of force. But don't overtighten and strip threads or pull out a torque wrench—use your common sense.

With every router you get one or two locking wrenches for the collet nut and shaft. For some reason, manufacturers try to skimp on these wrenches, and many of them are just thin pieces of poorly fitting steel. I replace my wrenches whenever possible with wider, better-fitting wrenches that I buy used.

Tighten or loosen the collet nut with the router unplugged and placed securely on a stable surface. I find it much simpler to have the tool resting on its side. Loosening the nut can present difficulties. If possible, find the wrench position on the nut that allows you to span both wrenches with your hands. Then pull them in toward one another with your fingers out of the way (see the photo on p. 27). When the nut finally breaks loose, you don't want your fingers catching the brunt of the force. If a nut doesn't want to loosen, put a cheater bar, such as the pipe from a pipe clamp, over the end of the wrench to improve your leverage.

To loosen a collet nut on a plunge router equipped with a spindle lock, use a slightly different approach. Spindle locks, which are found mainly on plunge routers, are buttons or levers that prevent the shaft from turning when engaged fully. You still use a wrench, but you have to make sure the wrench doesn't break free and damage a plunge column. Position it far from the column, or protect the column with a piece of scrap.

Removing stubborn bits

Many of us have had a bit stick in a collet or a collet stay jammed inside a shaft. There is only one safe approach. Loosen the nut a half-turn or so and support the shaft and nut on a bench top. Then holding a heavy wrench or other thin but heavy tool, rap the heck out of that nut, as shown in the photo at left. Make very sure you hit only the nut, never the threads. Leave the nut most of the way on so the threads are covered up, but make the nut loose enough so the collet can loosen.

If your first attempts don't free the bit, spin the shaft and try another flat on the nut. Keep rapping until the collet comes free. I have had one router for 15 years, and this is the only way to free up the bits. But this method always works, and I've replaced the bearings on the tool only once. It's a safe method if the shaft stays supported and you rap just on the nut. If you hit the bit, the carbide will break.

Some people resort to taking off the nut and hitting the collet, but that's not a good idea because it almost guarantees a dinged-up collet or shaft threads. Don't risk it. If you've pulled the bit up from the bottom of the shaft and can afford to lose some carbide, you could try banging on the end of the bit with a block of wood covering it. But this approach is for desperate circumstances only.

If your bits won't come out except with a block and tackle, replace your collet. Then check your bit shanks to make sure they haven't been damaged. Finally, check for damage to the motor shaft by inserting a new bit in a new collet.

CHAPTER THREE

MAKING ROUTER CUTS

Depth of cut is a critical setting. It should be determined by the hardness of the wood, the power of the router, and the sharpness of the bit. Usually it's better to reach full depth in several passes, rather than all at once.

Each woodworker has thresholds for pain, aggravation, and noise. One maker might find it a reasonable thing to plow a mortise in an inch of oak with one pass of his router, never giving a thought to the noise, dust, and smoke. He might also believe that replacing bits weekly and routers monthly is reasonable. I am happy to say I don't share shop space with this kind of maniac. My approach to cutting with the router is to keep the strain on both the tool and me to a minimum.

Router cuts should be made within the limits of the tool and the material. Some of these limits, such as horsepower, are out of your control once you own the router. Three important limits over which you do have control are depth of cut, feed direction, and feed rate.

DEPTH OF CUT

A lot of variables come into play when you determine the depth of cut. The material being cut can range in hardness from soft pine to hard maple and particleboard; an easy pass in one would produce bad results in another. Also, each router also has only so much power and a bit is only so sharp. It's difficult to give standards that will cover every situation. But here are some common-sense guidelines for topside and router-table work that will help, along with instructions for setting the depth on fixed-base and plunge routers.

Topside routing

It's easy to know when you're getting a clean cut made topside with a fixed-base router. The motor runs smoothly at a constant rate of speed, slowing only slightly as it enters the cut. The pass is made with a minimum of effort and noise. The results are clean and consistent throughout. A conservative approach to depth of cut will save wear and tear on your bits and bearings. In hardwoods, a 1/8-in.-deep pass will give you these results. In a softer material, such as pine, you could easily make a 1/4-in.-deep pass if your router has enough power (see the photo on p. 29).

A bad cut is just as easy to recognize. The router is hard to push through the cut, and the racket is excruciating. Dust is produced instead of shavings, or the aroma of charring wood fills your nose as the bit burns inside the cut. Fuzziness or chatter marks mar the new surface, and the depth of cut is not consistent.

These problems may not be related to taking too deep a cut; you also must consider the amperage of the motor, the sharpness of the bit, the hardness of the material being cut, and the size of the bit shank. But you will avoid a lot of problems by taking several light passes instead of one deep one. This is easier on your bit and collet because it reduces vibration and lessens the chance of chipping the carbide, burning a bit, or having the bit loosen in the collet.

Don't take a deep pass with an underpowered router—you'll hear it complain. Larger motors can handle a deeper cut, but as a bit's edge wears out, it will have more trouble getting through the work.

Bits should be kept clean to minimize further dulling; I soak mine in oven cleaner and clean them off with a toothbrush when they get encrusted.

Materials can vary widely in their abrasiveness. Cuts in composite materials dull bits quickly, especially high-speed-steel bits. And there can be variations in hardness within a species. If you cut into knots, your bit will take the brunt of it. Cuts on end grain are tougher than cuts in long grain, and there's also a greater chance of burning the wood and the bit. (A burned bit will get coated with resins and not cut as well.) Whenever possible, use bits with $\frac{1}{2}$-in.-dia. shanks. A $\frac{1}{4}$-in. shank bit will deflect and chatter more, especially in a deeper cut. If you must make a deep cut, make sure you're pushing a bit with a $\frac{1}{2}$-in. shank.

When using profile bits or pattern cutters, sometimes the only option is to set the bit at full depth. Try not to take a big bite all at once, but come into the cut slowly, and take several passes until the bearing finally comes into contact with the wood. You will be cutting freehand until this happens; only the last clean-up pass taken should be using the bearing.

For topside cuts with a plunge router, follow the same guidelines as for topside cuts with a fixed-base router. With a plunge router, having enough power isn't usually a problem. When cutting mortises, I rarely lock the router at a depth with the plunge lock except for the last pass. I just push the motor down a little bit and start to cut, making sure that a pass goes all the way to the ends of the mortise.

On a fixed-base router, bit depth can be checked with a ruler that has an end scale (top). You can also use a marked piece of wood to gauge the depth of cut (above) if the hole in the baseplate is too large for your ruler to span. Or you can make a notched depth gauge to fit over the bit for setting repeatable cuts.

There's a little surprise waiting for you in the form of bit chatter if the bit runs into a lot of wood at the end of a mortise.

Router-table cuts

I set my depths just the same on the router table as I do for topside cuts. A depth of 1/8 in. per pass doesn't put a strain on the 1½-hp motor I use in my table. Very often, for through cuts, I remove most of the waste first on the table saw with a series of quick passes. Then I set a bit to make one final cleanup pass. Also, when you feed past a bit, make sure that the stock is held down firmly to the table; do not allow it to ride up on a bit. Use featherboards when pieces need good pressure down to a bit or in to a fence.

Setting and adjusting the depth of cut

On a fixed-base router, depth of cut is set by moving the motor up or down in the base to expose the bit. Most fixed-base routers have some kind of scale on them for fine adjustments, but these aren't always accurate. So you need another way to set the depth. There are a few ways to do this.

Your tool kit should include some kind of small ruler. I use a 6-in. rule with several scales on it; the scale that I use most is in thirty-seconds of an inch. The best parts of this rule are the scales engraved on its ends; I can extend either side over the hole in my router base to measure bit heights (see the photos on p. 31).

Another way to set a bit is simply to mark a stick at the required depth and use that as a "ruler."

For bit settings that are used repeatedly for a project, you can make depth gauges of wood or plastic.

On a plunge router, depth of cut may be set in various ways. You can expose the bit at a final depth setting, just as you would on a fixed-base router. But if you do this, you can't take advantage of the plunge capacity, depth scale, and fine adjustments that have been designed into this tool.

Depth of cut is more easily set on the depth scale mounted on the outside of the plunge base (see the photo at left on the facing page). First, zero the bit on the work surface by plunging the bit down into contact with it, and lock the router in place. Then move the depth-stop rod down to one of the depth stops. Measure up on the depth scale the amount you want to cut, move the stop rod to this spot, and lock it in position. This will give you the travel and depth of cut you want.

On some routers, you can use an adjustable depth scale that can be moved exactly to a zero setting (see the middle photo on the facing page). Zero the bit on the work and dial the scale in to 0, and then dial the correct depth in on that scale.

On plunge routers, fine adjustments are simpler to make and more accurate than on fixed-base routers. If you rotate the turret stop in place under the depth-stop rod, several different depths of cut can be made with one bit. Some turret stops also have threaded screws in their ends that can be moved in or out for

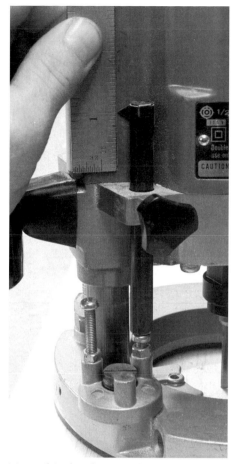

To set bit depth on a plunge router with a fixed scale, zero the bit on a flat surface and drop the depth rod down to a turret stop. Align a ruler with the pointer of the depth rod, read up on the ruler the amount you want to cut, move the pointer to that mark, and lock the setting in place.

To set bit depth on a plunge router with a movable scale, zero the bit on the work surface and move the depth rod down to a turret stop. Then align the pointer with zero on the scale, and dial in the depth of cut from this point on the scale.

There can be several fine adjustments to a depth setting on a plunge router. These include a rotating turret stop, turret screw adjustments, a screw adjustment to the depth rod, and add-on adjustment screws to the height rod.

fine corrections. The depth-stop rod may have an adjustment screw on its end for fine tuning (see the photo at far right above). Some routers have a built-in fine adjustment screw on their height rod, while most others can have one added onto it to move the router small amounts.

There are also situations where the upward travel of the bit needs to be restrained as well as the downward travel. For example, allowing a dovetail bit to lift up

could ruin a cut or a jig. Most plunge routers have a set of nuts on the height rod to prevent unwanted upward travel. Better-designed routers have some kind of quick release to speed up this adjustment.

After setting the depth of cut on the depth scale, make a practice pass in a piece of scrap. It's always a valuable use of your time to double-check your settings before committing your good wood to a cut.

TOPSIDE CUTTING: FEED DIRECTION

Move the router so the wood is moving into the rotation of the bit.

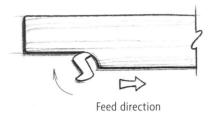

Feed direction

ALONG ONE EDGE OF A BOARD

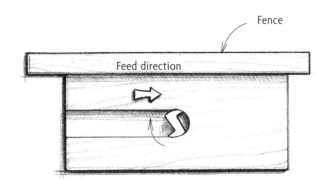

Fence

Feed direction

**IN THE CENTER OF A BOARD,
WITH A FENCE**

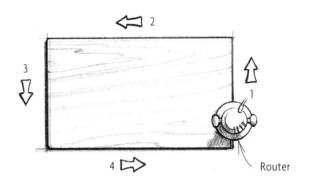

Router

A cut made around the edges of a board should be made in a counterclockwise direction. Start on an end-grain surface and follow with a long-grain pass to clean up any tearout that may occur.

AROUND THE EDGES OF A BOARD

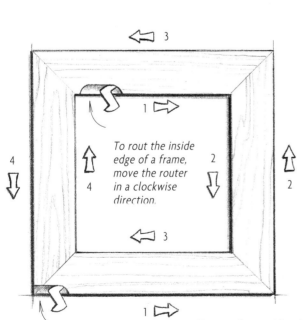

To rout the inside edge of a frame, move the router in a clockwise direction.

To rout the outside edge of a frame, move the router in a counterclockwise direction.

AROUND THE EDGES OF A FRAME

FEED DIRECTION

Because a router bit does most of its cutting on its edge rather than straight down like a drill bit, feed direction is important. The direction in which you feed the bit can help to create—or destroy—a joint. Held topside on a board, a router bit spins clockwise as you look down on it. Mounted upside-down in a router table, a router bit spins counterclockwise as you look down on it. Depending on how you move the router (topside) or the workpiece (router table), the bit will either pull itself into the cut, as it usually should, or push itself away.

Let's talk first about the rules for feed direction. Then I'll tell you when you can break them.

Topside routing

Hand-held on top of a board, the router should be moved so the wood is feeding into the rotation of the bit (see the drawing on the facing page). As it cuts, the bit will pull itself into the workpiece. When working on the edge of a board, move the router left to right. When working on a frame, move left to right (counter-clockwise) around the outside of the frame. On the inside of the frame, move the router in a clockwise direction. With a bearing-mounted bit this feed direction will pull the bit into the wood right up to the edge of the bearing.

When using a straight fence (see pp. 44-48) to cut along the edge of a board, feed in the exact same direction: left to right. This will pull the fence in tight to the edge. When making a cut set in from the edge, this feed direction, against

the rotation of the bit, will do the same job of pulling the fence in tight to the edge.

When running the router base or a template guide against a fence, the fence can be placed on either side of the router, so it's tricky to decide which way to feed. Pay attention to the bit rotation as your key. Always move the router so the bit pulls the router into the fence. This feed direction is opposite the bit rotation.

What if you disregard these guidelines and feed with the rotation of the bit? You'll feel the bit try to push itself and the fence away from the cut or skitter along an edge. If you cut in this direction, you will have to work to hold the fence in tight to the edge.

Router-table cuts

When you flip the router over to cut in a router table, the bit spins counterclockwise as you look at it from above, so you have to reverse your feed direction (see the drawing on p. 36). On the router table, you feed the work into the bit from right to left, into the rotation of the bit, to push the workpiece into the fence during the pass or to pull the wood into the edge of the bearing. As you stand at the front of the table, the first contact the bit makes with the wood is at the front of the bit's cutting edge.

If the first contact is at the back side of the bit, however, feeding from right to left means feeding with the rotation of the bit, not against it. On the router table, this can do one of two things. It can scare the heck out of you as the

ROUTER TABLE: FEED DIRECTION

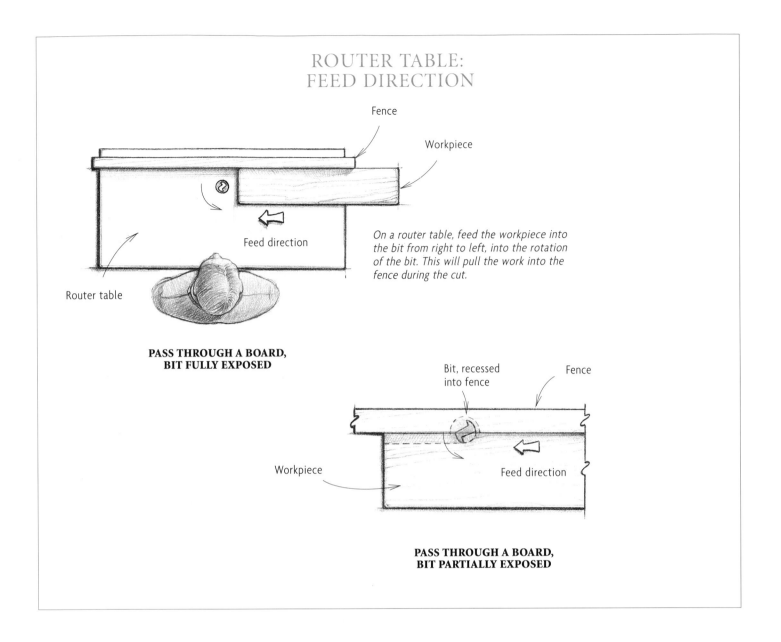

Fence

Workpiece

Feed direction

Router table

On a router table, feed the workpiece into the bit from right to left, into the rotation of the bit. This will pull the work into the fence during the cut.

PASS THROUGH A BOARD, BIT FULLY EXPOSED

Bit, recessed into fence

Fence

Workpiece

Feed direction

PASS THROUGH A BOARD, BIT PARTIALLY EXPOSED

piece of wood shoots out of your hand following the bit's lead, or you might manage to hold onto it if you fight to keep a grip. Either way, it's dangerous.

Always be sure you know which side of the bit will be doing the cutting in a pass, especially when routing small pieces. Sometimes a first cut is made, and you follow that with a second pass. This can happen when a straight bit is too small for a cut. If you move the

fence so that the far side of the bit is cutting, you will be feeding with the rotation of the bit, and you could lose your hold on the workpiece.

When using bearing-mounted bits on the router table, be careful when starting a cut. Many commercial router tables use a starting post as a pivot point to begin a cut. Simple care instead will allow you to begin a cut safely without using the pivot point. Start your cut just back from the end of a board to prevent an unwanted cut or tearout, because it's difficult to hang onto a piece if it starts to run backwards. If the bit catches the end grain as you begin the cut, it can take and kick the end of the board past the bit (see the photo at right). For added safety, use a fence with a bit recess so there's never a chance of catching the end grain.

Starting a cut on the router table can be tricky. If you catch the end grain with a cut, the bit will shoot the board past the bit and run along the end grain, ruining the cut (left). Start the cut just in from the end to prevent this (right), or use a fence with a bit recess or a pivot post on the table to help start the cut.

Climb cutting

Often a router cut is made properly, but the results are all wrong. The wood tears out ahead of the cut or chips on its edge. Why? When cuts are made in the proper feed direction, there is no wood supporting the last fibers as the bit emerges from the cut. If the grain direction of the wood opposes the cut, the wood will tear out ahead of it.

Now it's time for rule breaking. Because no matter what this or any book will tell you, the best advice is always to do what the wood tells you to do. As long as you don't endanger yourself, forget the rules if they don't help.

Climb cutting, or cutting with the direction of bit rotation, avoids mishaps by scoring the wood first with a backwards pass. When you move a router backwards through a cut, the router will scoot right along the edge—you actually have to restrain the router to keep it from moving along too quickly. But what climb cutting will do is make cuts into the wood that are always supported by other fibers. Then you can come back and cut to depth in the proper feed direction with no danger of tearout (see the drawing on p. 38).

Climb cutting can be used effectively for freehand routing. When I rout for inlays or hinges, I often rout out most of the waste. I always take light passes when working freehand, and feed the bit so it pushes itself away from the wood. This gives a more controlled pass. You work to keep the bit in the cut and nibble away at the wood rather than fighting to keep the cut consistent without the bit pulling itself in too deep.

A flush-trimming bit is usually set to cut at a full depth. This can be hard on a bit in a deep pass. If you run the bit backwards along an edge, then it will just nibble away at the wood without biting in too deep. Several freehand passes will get the wood close enough to the pattern that a final pass can be made in the proper feed direction.

CLIMB CUTTING TO PREVENT TEAROUT

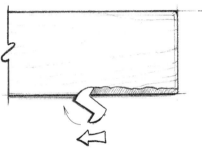

Feed direction

A normal feed direction can sometimes produce tearout on the unsupported fibers at the edge, especially when cutting against the grain. Tearout can also occur below the final depth of cut. (Bit rotation is clockwise on a topside cut.)

Feed direction

A climb cut scores the edge as the bit tries to climb out of the cut and push itself away from the edge.

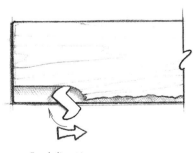

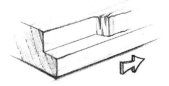

Feed direction

A cleanup pass to full depth will take care of any further tearout. Any tearout that occurs will be set back from the edge.

On the router table, climb cutting has to be done with great care. Use the fence as a guide for this cut, and always keep your fingers out of harm's way. It's easy for the bit to grab the workpiece and pull it into the bit, along with your fingers! Exercise great caution.

FEED RATE

Every router has a characteristic howl to it. If it's a variable-speed motor the speed and sound should stay constant, even under a load. But with any other type of router motor, you will hear the motor first get up to speed and then slow as you enter a cut. You may need to adjust your feed rate accordingly.

Bogging vs. burning

Learn to recognize the normal sound of your motor under a load. Always pay attention to it as an indicator of how the cut is going. When you're moving the router or the workpiece at a proper feed rate, the sound of the motor should remain steady. When the feed rate is too fast, you'll hear the motor bog down or the router bit scream as you try to make the cut. Or if a bit slips and starts to work its way out of a collet, the sound of it taking too deep a bite will tell you something is amiss.

When the feed rate is too slow, the wood and the bit will start to burn. In end grain, which loses moisture quickly and is so much harder to cut than long grain, burning occurs quickly when the feed rate is too slow. Add a dull bit to a slow feed

A burned bit (the bit on the right) will not stay sharp or cut cleanly. Burned bits should be cleaned to increase their life, or sharpened by a professional.

rate, and you'll have a shop filled with the aroma of burning wood. Try to move through end grain at a fair pace because a slow hand here will give you more to clean up. Burning will also occur any time you stop a cut with the bit touching the wood, because the bit is just spinning in one spot and building up heat (see the photo above).

Feed rates can also determine the quality of a cut. If the rate is too fast, the bit may chatter, yielding a poor cut. Tearout may also occur. Listen carefully as you make a cut. If you hear the wood splinter, slow down your feed rate.

Pilot-shaft and bearing marks
High-speed-steel bits with pilot shafts (see p. 20) are notorious for leaving their traces behind. All that friction builds up a lot of heat in a hurry, which causes a burn mark all along the edge of a cut. Try to take several freehand passes before getting down to a final cut so the pilot doesn't contact the work until the very last pass. If you're taking only a little off with each

cut, you'll be able to move the bit just a little more quickly and cut down on some of that burning. Keep the pilot shaft as clean as possible with oven cleaner or lacquer thinner.

Bearings should be less of a problem. However some composite bearings will always leave behind a burn mark right where the bearing first comes into contact with the wood. Steel bearings won't have this problem as long as they're spinning properly. Check your bearings to see that they move with no awful crunching sound; a little noise won't affect their movement. And check to see that no debris has gotten caught around a bearing. This can make it difficult to turn as well.

CHAPTER FOUR

GUIDING THE CUT

All these bases can be used to help guide a cut. The plunge router at left has a large, fully round base; the plunge router at right has a straight edge along one part of its base. The fixed-base router at center has a large Plexiglas sub-base with a straight edge cut into it.

I have a friend, a rocket scientist (I kid you not), who once described the router as the quickest way to ruin a piece of wood—he said he could ruin it more slowly with hand tools. And there's a lot of truth to that tongue-in-cheek statement. People have the mistaken notion about routers that if you can just figure out how to get a bit into them, everything else will fall into place. But like all the other tools in the shop, you need a careful way of directing the router to produce the results you're after.

With power tools, such as the table saw or jointer, or hand tools, such as the hand plane or chisel, you don't just plow into the wood hoping to get perfect results. You reference off straight and flat surfaces that place the cutting edge of the tool in a position to make a controlled cut parallel to that reference surface. For example, in a table saw, the table and fence offer two surfaces to reference a cut from. The hand plane uses the sole of the plane as a reference surface, while the chisel uses the back or bevel of the tool.

Similarly, the router is a useless tool until you tell it where to cut. Place a bit in a router with no guide on it, and you'll end up routing every which way except the one you need. By using reference surfaces as guides, a limitless number of cuts can be made safely and accurately, producing predictable results.

There are many clear and simple methods of guiding the router, which is why there are so many jigs, fixtures, and gadgets on the market. We have already seen how some types of bits use bearings or

pilot shafts to guide the cut (see pp. 20-22). In this chapter we will take a look at four other ways to guide a router: router bases, fences, template guides and templates, and router tables.

ROUTER BASES

A router base may look like just a way of placing the motor and bit down onto a surface, but it is in fact also the easiest way to guide a cut. Once the bit height is locked into place, the base edge is fixed relative to the bit, and therefore can act as a constant referencing guide from the cutting edge (see the drawing below). What this means is that if you run one edge of the base against a fence, and

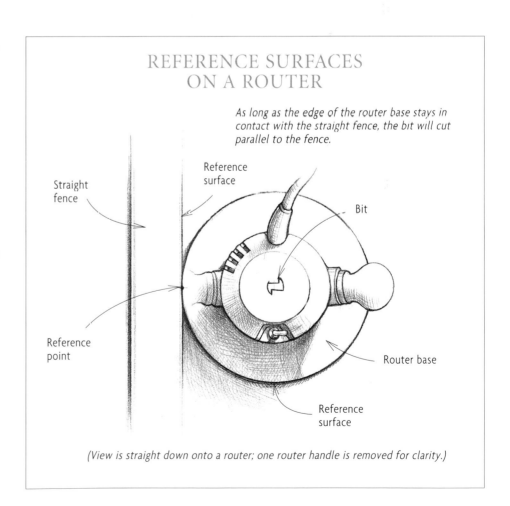

REFERENCE SURFACES ON A ROUTER

As long as the edge of the router base stays in contact with the straight fence, the bit will cut parallel to the fence.

Straight fence

Reference surface

Bit

Reference point

Reference surface

Router base

Reference surface

(View is straight down onto a router; one router handle is removed for clarity.)

you are vigilant in keeping that contact point, you will always get a cut that is parallel to that fence. Always.

With a router base as a referencing guide, the tool can be run against many other surfaces to produce cuts. These can include both shopmade and commercial straightedges, jigs, and fixtures. With known bit diameters, you can measure the distance from bit edge to base edge and use that number to set up fences for a variety of cuts.

Checking the base and sub-base

In any router that you hope to use accurately, the bottom of the base and sub-base must be flat. Metal bases are machined flat at the factory, so if flatness is a problem it's usually in the plastic or phenolic sub-base. Check them both on a flat surface, such as the top of a table saw.

Check the sub-base with a straightedge for high spots or distortions. If it's not flat, remove the screws that hold it in place and make sure there's no dust or debris between the metal base and sub-base that would prevent a flat fit. Clean off the underside of the base occasionally with compressed air or a brush to prevent the buildup of dust. (Sub-bases can also be sanded flat on the belt sander.) Sub-bases can warp, but they're easily replaced. Make certain the screws that hold the sub-base in place are not distorting it by being overtightened.

Check the base, making sure it doesn't rock, and look for high spots. If there is major distortion

on a new base, have it replaced. Small high spots can be rubbed out on a piece of silicon-carbide sandpaper set on a flat piece of glass.

Router bases are usually fully round in shape but some are partially round with one straight edge built into them. The straight edge on a base provides another reference surface that you can use for guiding your cuts.

Keep in mind that round bases may not be concentric with the router bit or not perfectly round; one side may be slightly closer to or farther away from the bit. For this reason, whenever you use a round base to index a cut, always use just one point on the base to reference from (see the photo on the facing page). In other words, do not rotate the router base as you make the cut. Some makers put a mark on their base as a guide. I use a point directly under one of my handles as my reference point.

Auxiliary sub-bases

An auxiliary sub-base can provide even greater stability and versatility for the tool. For my fixed-base router I made an auxiliary sub-base out of a piece of 1/4-in. acrylic plastic. This 9-in.-dia. base gives me a wider platform, which is useful when I'm resting the router on a narrow surface. I've also machined a straight edge into this base so I can use it against a fence (see the photo on p. 40). Offset bases with a teardrop shape and an extra handle can also be used to provide added support for some cuts.

Longer straight bits will often expose too much cutting edge in a fixed-base router even when the

To maintain accuracy, keep a one-point contact with the base right underneath the handle against a straight fence. Or mark a spot on the base. Just don't rotate the base as the cut is made. In this setup, the ¾-in. plywood spacer between the base and sub-base allows a longer bit to be used in this fixed-base router.

motor is drawn up to its highest point. The problem is easily remedied by adding a spacer between the base and sub-base. The spacer should be cut at the same diameter as the base and sub-base and drilled with mounting holes for screws. It will give the router a longer dimension, so less bit will be showing.

FENCES

Any straight edge that the router base can index or reference its cut from is called a fence. Fences may be shopmade or purchased. The simplest type that you can make, used for topside routing, is a straight-edged board clamped onto your workpiece. The router base runs against this straight edge with the bit away from it at some measured distance. That's all there is to it.

Shopmade fences

The right-angle jig, also used for many situations in topside routing, is a straightforward fence to make, but it requires accurate construction in order to be worth the scrap it's made out of.

This jig is actually made up of two fences connected together (see the photo on p. 44). The upper piece is a flat and square plate; the lower piece is a straightedge glued to it at exactly 90°. (If you make a rabbeted cut into the upper plate, this will help locate the fence at 90°.) When this jig is placed tight against the edge of a workpiece, the straight edge of the plate is automatically set at a right angle to the edge of the workpiece. The router base runs along the edge of the plate, producing a cut that is square to the edge of the workpiece.

RIGHT-ANGLE JIG

The dimensions can be changed to meet other shop requirements. But the pieces must be assembled at exactly 90°.

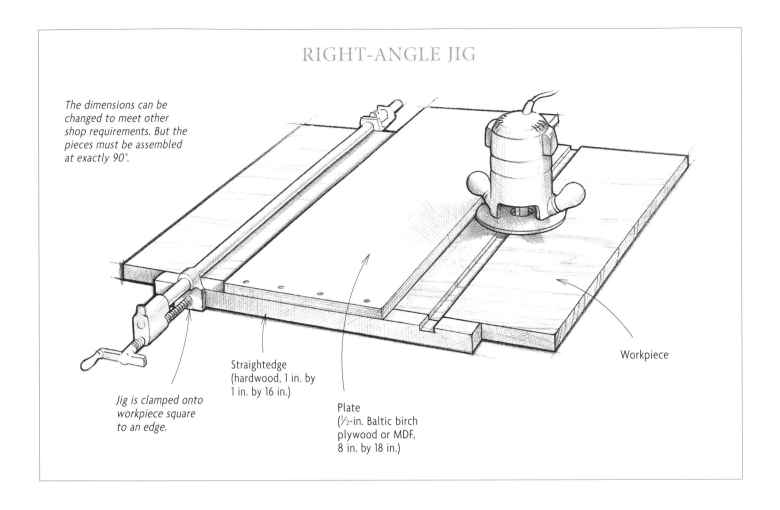

Jig is clamped onto workpiece square to an edge.

Straightedge
(hardwood, 1 in. by
1 in. by 16 in.)

Plate
(½-in. Baltic birch
plywood or MDF,
8 in. by 18 in.)

Workpiece

I use my right-angle jig when I'm cutting topside dadoes on pieces that are too large or cumbersome for my router table. It's also useful when cutting stopped dadoes or sliding dovetails because it's easier to see where the cut should stop than on a router table.

The jig can be used with a router in two ways. The router base can ride against the plate as you hold one point of the base against it; the router base rests on the work surface being cut. Or you can use a template guide in the router (see p. 48) with the router resting above the workpiece and on top of the plate.

With either method, you have to determine the offset of the bit's cutting edge to the locating point. If you always use the same router and bit with your jig, you can measure this offset easily and mark it right on the jig. If you use a template guide for various-sized bits, you can measure those offsets too, and mark them on the jig for easy referencing.

Straight fences

If you could add something to the router base that would provide another reference surface and also be adjustable relative to the bit, the router could make cuts on an edge with greater ease and accuracy. This is exactly what a straight fence does.

For dadoes, grooves, and mortises cut topside, a router and straight bit need some kind of reference guide. Mounting a fence to the router offers an adjustable method of indexing the cut. Instead of the fixed distance from bit to base edge, the fence distance can be varied. The router and bit can be adjusted so that any distance from 0 to the length of the rods on the fence can be set. This makes setting up for cuts much easier than indexing off the base edge when routing topside. So how do these fences work?

There are plenty of different designs (see the photo below), but they share these general features. Two steel rods are connected to a straightedge. The rods fit into holes in the router base and are locked into place with screws or wing nuts. The straightedge body is usually a yoke shape ending with two short lengths of steel that provide the guiding surface.

The fence is adjusted in one of several ways. The straightedge can move freely on the guide rods into position and then be locked into place on them. Or it can be permanently fixed to the rods with the whole unit moving in and out relative to the router base. Screws or wing nuts lock the rods—and thus the fence—in position.

The problem with most fences is how poorly they adjust and hold their settings. My first router came with a fence whose fine adjustment was a marvel of bad engineering. The guide rods were difficult to lock into the base. There were impossibly small reference surfaces on the straightedge, and the screw adjustment was horribly sloppy. The whole thing was useless.

Make certain that your own fence mounts on the router easily and locks down tightly and accurately. It should adjust smoothly and have little side-to-side slop.

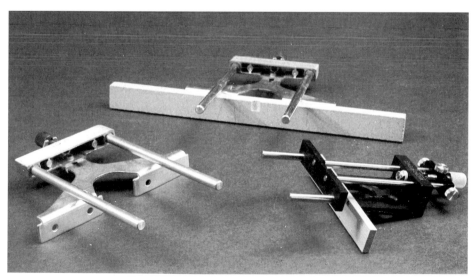

On straight fences, steel rods mount to the router; the location of the straightedge is adjustable. The fence in the middle of the photo has a long wooden auxiliary fence screwed onto it. There is also a recess routed into it for the bit. The fence on the right has a micrometer adjustment for precise movement.

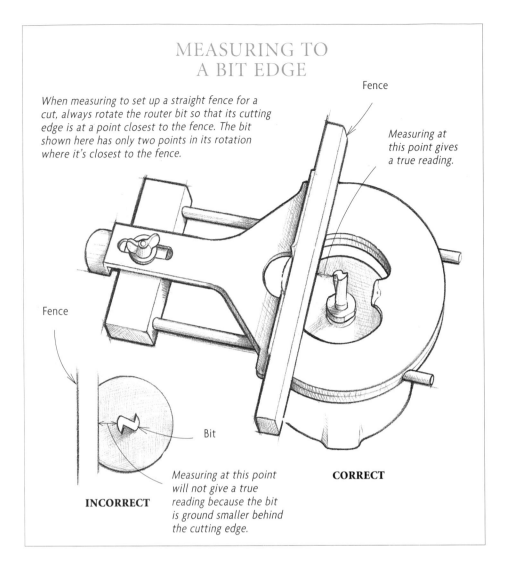

MEASURING TO A BIT EDGE

When measuring to set up a straight fence for a cut, always rotate the router bit so that its cutting edge is at a point closest to the fence. The bit shown here has only two points in its rotation where it's closest to the fence.

Fence

Measuring at this point gives a true reading.

Fence

Bit

Measuring at this point will not give a true reading because the bit is ground smaller behind the cutting edge.

INCORRECT

CORRECT

Most straight fences also have some kind of fine-adjustment mechanism for moving the straightedge very small amounts. Generally a lock screw or wing nut is loosened, then the fine-adjustment knob can be turned to move the fence to the exact position required.

Another method of setting a straight fence is to mark out the wood to be cut and to bring the router fence and bit up to it. (The bit will have to be rotated so that the bit edge is closest to the fence.) Once you set the bit onto the work, this kind of absolute evidence will tell you immediately if the fence is or is not in the proper spot.

A really simple improvement to a straight fence is to screw on a longer auxiliary fence made of a straight piece of solid wood or plywood (see the photo on p. 45). Long auxiliary fences provide greater stability and accuracy for all cuts, especially cuts near the ends of boards, where a heavy router and a short fence would tend to pivot off the edge of the board. For cuts with only part of the bit exposed, the fence can also have a recess in it for the bit to hide in. To create this pocket, the fence is mounted onto the router and the bit slowly plunged into the wood fence.

Another useful addition is to clamp a secondary fence to the router base with small C-clamps to hold the router in place on an edge (see the photo on the facing page). Make sure the fence is deep enough and long enough to

When adjusting a fence distance, always make sure the bit is rotated so that one cutting edge is at a point closest to the fence. This is a small but critical thing. As you look at a two-fluted straight bit, it's apparent that the cutting edges do not take up the full circumference of the bit. They rotate throughout the cut and are closest to the fence at only two points (see the drawing above). With the router unplugged, you need to rotate the bit to move it so one edge is closest to the fence, then measure for fence distance.

contact the work securely. Set the adjustable fence first and place the router onto the work. Then clamp on the secondary fence, making sure the clamp handles don't get in the way of anything.

One last warning about fences: With a straight fence, take care when feeding the router into the work. A fence is no guarantee the cut will be made parallel to the edge. If you let your mind or your grip wander, your cut will too. Move the router into the work so that the fence will always be pulled into the work. A left-to-right feed direction when routing topside (see pp. 34-35) ensures this.

A secondary fence clamped onto the router base will lock a base in position and help to prevent the router from tipping.

TEMPLATES

Templates are reproducible patterns that you can index cuts from. If the pattern is made precisely and the router follows it accurately, the results will always be predictable.

The possibilities for shop-built templates are almost limitless. Think of a shape and make a pattern of it, and you've got a template. Templates can be made out of a variety of stable materials, such as plywood, MDF, Masonite, plastics, or aluminum, and used until they get nicked by a bit or wear out.

Templates can be used in one of two ways: with a flush-trimming bit or with a template guide. With only a flush-trimming bit, an exact duplicate of the template shape will be produced, since the bearing and cutting edge of the bit line up. Template guides (see the sidebar on p. 48) will also follow a

template, but their cut will be offset by the distance between the guide's outer edge and the bit.

Templates with flush-trimming bits

When using a flush-trimming bit with a template, always make sure to trim the workpiece first on the bandsaw or with a jigsaw close to the final shape. You should leave only ⅛ in. to ¹⁄₁₆ in. of waste. This way the bit doesn't have to work so hard, especially when cutting in thick or hard stock. Nibbling passes can be made with the flush-trimming bit when necessary if too much wood is left to cut in one pass (see pp. 37-38).

Also watch for cuts against the grain. There is no guarantee the wood will nicely trim up flush in all directions. If the wood looks as if it

TEMPLATE GUIDES

Template guides are round metal plates that fit into the bottom of the router base. As seen in the drawing below, they have a thin-walled collar that extends from their base. A router bit fits through the collar without touching the inside wall. The outer wall rubs against a straightedge or template and guides the router to produce a cut.

The advantage template guides provide is a greater degree of accuracy because the offset from the bit's edge to the guide's outside wall is very small. The bit will be equidistant, or very close to equidistant, from all points on the guide's outer walls. This is especially apparent when compared to the distance from bit edge to the edge of the router base or straight fence. Template guides are mounted firmly in place, so there's no slop in them like some fences. Finally, they can be used with templates to reproduce both simple and complex shapes.

Because of the different bit sizes and variety of templates, there are many guide sizes. Router manufacturers usually provide some type of template guide to fit their routers. Most manufacturers also provide adapter collars that mount onto the router base for use with screw-on standard-sized template guides. These guides come in a range of diameters for different applications. To ensure the accurate placement of a guide after a sub-base has been removed, check its location relative to a mounted bit to see that the guide will be concentric to it.

might give you trouble, take a light pass to see how it cuts. If the piece has too steep a curve, go only halfway down the curve, flip the template to the other side of the work, and finish the cut in the proper direction. Templates can be held in place on workpieces with small brads, screws, quick-action clamps, or double-stick tape.

Templates with template guides

When template guides are used with a template, there is always an offset from the guide to the bit edge. (The router bit has to be small enough to fit through the template guide, after all.) The first thing to determine is what this offset distance is, so you can allow for it in your setups. To do this measure the diameter of the template guide with a set of calipers and subtract the diameter of the router bit, then divide by 2.

Another method is to mount the bit and template guide in the router, plunge the bit through the guide and measure from the cutting edge of the bit to the outer edge of the template guide.

In some situations (e.g., a dovetail template), this offset measurement is not critical to know. If you use the proper template guide for a jig with a specific outside diameter, or O.D., the most crucial dimension is the bit height. However, when making a template for routing mortises (see the drawing on the facing page), the offset dimension is important to know so the mortises come out properly sized.

Mortising templates can fill with debris as cuts are being made, so always make a final clean-up pass

Router base

Template guide

Bit

Collar

Template guide

Template guides are fixed into the bottom of a router base. They have an inside and outside diameter. A router bit should spin freely inside a guide.

A TEMPLATE GUIDE
WITH A MORTISING TEMPLATE

A template slot is cut larger than the mortise to be cut because of the offset between the bit's edge and the outside diameter of the template guide. This offset has to be figured into the template shape to produce a properly sized mortise.

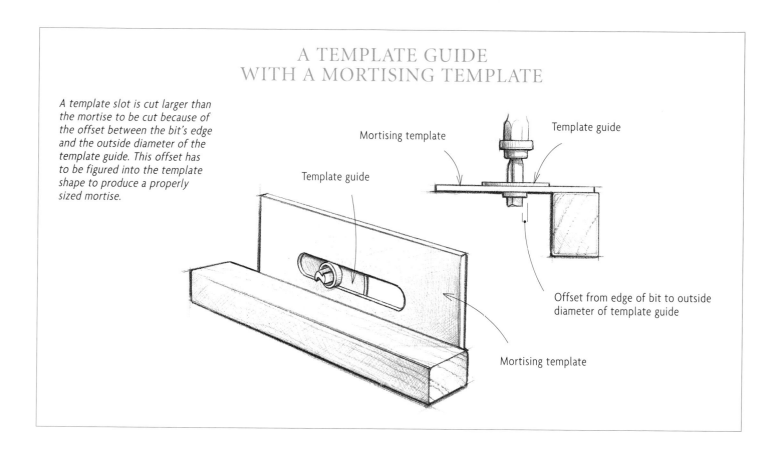

Mortising template

Template guide

Template guide

Offset from edge of bit to outside diameter of template guide

Mortising template

---

after clearing away all the waste from inside a template. Compressed air or a pencil works fine to clean out the ends of the template.

ROUTER TABLES

Some accessories do a marvelous job of extending the usefulness of the router, but others are foisted upon buyers with more dollars than sense. Router tables are one of those items where you can spend more time and money setting up the tool than working with it. I don't get misty eyed over the beauty of a router-table design. I use it to do work.

Router tables offer a completely different way of using the router. Instead of holding the router in your hands and moving it around a

workpiece, you mount it upside-down in a table and move the workpiece past a bearing-guided bit or a bit and fence. Stops can be mounted onto the fence for indexing cuts, and smaller pieces can be more safely routed than with a hand-held router. I do so much work on my router table that I cannot conceive of shop life without it.

Shop-built router tables

The first router table I ever saw was in California at the shop of Art Espenet Carpenter, one of the masters of the craft. It was just a small sheet of plywood with a router mounted underneath, and the whole thing sat over a 55-gal. drum. The drum caught the sawdust but also produced the loudest router noise I've ever heard.

A fixed-base router mounted to the underside of a 24-in. by 36-in. piece of melamine clamped onto the edge of a bench makes a simple but effective router table at a comfortable 38-in. height.

FENCE PLACEMENT ON THE ROUTER TABLE

A fence does not need to be square or parallel to any other router-table edge. The only important distance (X) is from the cutting edge of the bit to the face of the fence.

Fence

X

Router table

Router table

Fence

X

X

X

Fence

Fence

My own version is a piece of ¾-in. melamine clamped onto the edge of my bench top with a fixed-base router base mounted underneath it (see the photo at left). Plywood covered on both sides with laminate or MDF could also be used. I have a separate base for this router so that I can use it for topside routing as well.

I use several fences with this router table, but they're all pretty much the same: a straight piece of wood with one square face and edge that I clamp down to the table with two C-clamps. It's that simple. Some of the fences are taller for better support on wider stock, while others have cutouts for a bit to be partially hidden within.

Router-table fences do not need to line up with an edge of the table. They do not have to be parallel to a miter-gauge slot or run in a track. They can spin 360° around a bit because the only important measurement is from the bit edge to the face of the fence (see the drawing at left). If that distance is correct, then the fence can be in any position and still function properly.

Whenever a fine adjustment to the fence is needed, I use my patented fine-adjustment system. I pencil-mark the table right up against the fence, loosen one clamp, then with my hand or a hammer, tap the fence in the proper direction and lock it in place again. Since the fence pivots off the other clamp, if the pencil mark is made at the end of the fence it gives no clear indication of how much the fence has moved right near the bit. So make the pencil marks near the bit, where you can see some results.

Fences can also be marked to indicate the cutting edges of the bit. This is useful when plunging down onto the bit. It is also very useful when setting up stops. By marking the bit edges onto the fence, stops can be measured out and clamped into place on the fence. Before marking, make sure to rotate the bit so one cutting edge will be closest to a stop.

Featherboards are another set of hands in the shop, and who hasn't needed that a time or two? Cut on the bandsaw at an angle, with flexible fingers to provide good pressure against a fence or down to the table itself, these boards can go a long ways toward keeping the blood pressure down. Just make sure they don't impede your feed rate by being too tight. Pushing against the featherboard itself rather than the stock will provide good pressure at a safe distance for your hand.

There are other simple versions of shop-made router tables. Probably the most common is the use of a table-saw extension table as a router-table surface (see the photo above right). The advantages to this arrangement are a large work surface and easy access to the table saw fence. The disadvantage is that the table saw might not be usable if it's tied up as a router table, and vice-versa.

Another disadvantage of the table-saw router table is its height. As my knees and back get creakier, the less I like having to bend. If you have similar problems, you might find a router table like the one shown in the photo at right more to your liking. It's a four-sided box that can rest on a bench, sawhorses, or another table.

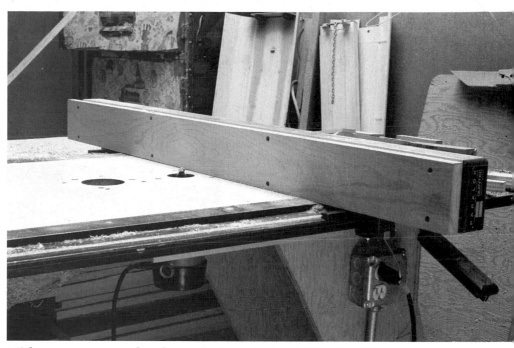

With a router mounted underneath, the extension table of a table saw can double as a router table. The 34-in. table height may be somewhat low for comfortable work, but the saw fence is handy and if easy to adjust, a real help when doing fine work.

This router-table box rests on a cabinet to bring it up to a 43-in. height. It also has a side mount for routing horizontally.

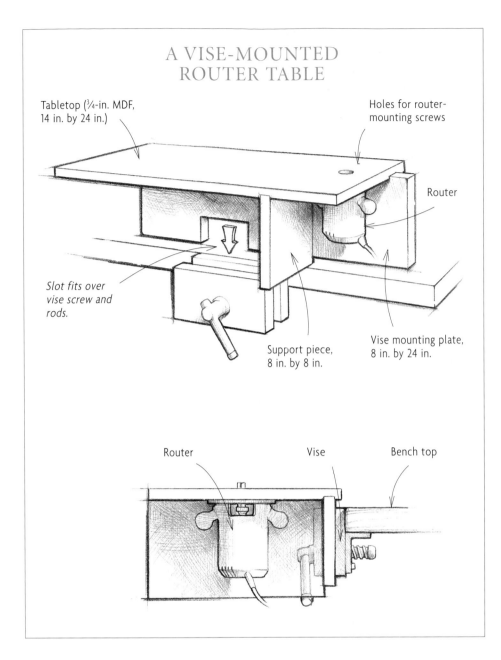

A VISE-MOUNTED ROUTER TABLE

Tabletop (¾-in. MDF, 14 in. by 24 in.)

Holes for router-mounting screws

Router

Slot fits over vise screw and rods.

Support piece, 8 in. by 8 in.

Vise mounting plate, 8 in. by 24 in.

Router

Vise

Bench top

It's made of melamine with a miter-gauge slot routed into the table surface. A tall fence clamps onto the table, which stands 43 in. off the floor. There is also a mounting plate on the side for mounting a router horizontally.

What happens when you have your router table set up for one operation and you need to make a different router cut? The

woodworker who made the router table shown in the top photo on p. 51 ran into this very problem, and his solution was the simple vise-mounted table shown in the drawing at left. Three pieces of MDF are glued and screwed together at right angles with the router mounted underneath.

Commercial tables and fences

There are many types of commercially available router tables made of a variety of materials. These include cast iron, cast aluminum, steel, and melamine-covered surfaces. There are also many separate fences and drop-in bases (also called table inserts) that you can buy. They too come with adjustments, attachments, and, of course, price tags. If you're in the market for a store-bought table or fence, here are some important things to look for.

A router-table surface must be flat and not twisted. Use a straightedge to check for flatness, and put a set of winding sticks on it to check for any twist. The table must also easily accommodate your router and base. If you use a plunge router and have to drop it into place with a plastic or phenolic base on it, make sure you can install and remove it with ease. You have to take out the router to change bits, so it's a movement you'll repeat.

More important, make sure your drop-in base or the one for the router table can be adjusted to be perfectly flush with the surface of the router table. It makes absolutely no sense to spend $100 or more on a router table and then have to build in a fudge factor with each cut.

Drop-in bases should also be made of phenolic for rigidity. More resilient plastics will bend under the weight of the router, thwarting your efforts to do precise work. If there are filler rings to cover the bit hole, make sure they fit in tightly with no danger of vibrating loose. Also be sure they line up or can be adjusted to line up perfectly with the table top.

If you use a plunge router in the router table, a height-adjustment knob is essential to make bit adjusting simpler. A knob can be bought and screwed onto the height column if your router doesn't already have one.

Commercial router tables and drop-in bases generally also have a pivot post to help start cuts made with bearing-mounted bits. The post helps to prevent the cutter from diving inadvertently into the end grain of a workpiece or to help support a cut in from the end of a board.

All commercial router tables and plastic drop-in bases have a problem, I believe, if used with a plunge router. A plunge router is usually not a light tool. Trying to drop that monster into a router table whenever a bit is changed is too much effort. For most of my topside routing except mortising I prefer a router I can handle with ease. The plunge router with a huge phenolic base on it just becomes too cumbersome for my purposes and difficult to use topside. The table-insert rabbet and drop-in base are also subject to wear and tear and thus accuracy problems over time.

For these reasons I use a fixed-base router in my shop-built router table with a base permanently mounted. There's no sag to worry about, it's simple to change bits, and it's only slightly more difficult to adjust bit heights. The horsepower of a fixed-base router is more than sufficient for my routing work. If I need to cover the bit hole for better support, I clamp a flat piece of ¼-in. MDF or Masonite over it with a small hole drilled into it for the bit to poke through. It's also simple to pull the motor and put it into a free router base for topside work. This smaller router is also much easier to hang onto for most topside work, and accuracy with a router can depend as much upon your grip as your measurements. When the table wears out, it's also easy and cheap to replace.

Fences are another item that can be simple or complicated. If you buy a fence, check it to see that it's straight and square to the router table. There's no point in having to shim a store-bought fence to bring it into true. Aluminum fences are available with dust port fittings, plastic bit guards, and adjustable split fences. These fences can be adjusted to show more or less of the bit, which can give better support to the workpiece. With wood dust becoming a greater concern, a vacuum pickup seems desirable, especially for small shops.

PART TWO

INTRODUCTION TO JOINERY

CHAPTER FIVE

THE MECHANICS
OF JOINERY

This cherry library table, designed and built by the author for the Oregon State Archives, shows the structural framework that joinery can create. Photo by Harold Wood.

Joinery is the backbone of furniture making. Around its central framework, furniture of strength, utility, and lasting beauty can be created. Joints provide structure for a piece, while also helping in assembly and influencing furniture design. Yet they are made out of wood, a material that continues to move about long after it's been milled into planks.

Over time, woodworkers have designed an astonishing variety of joints. A few of these joints use fasteners, but fine furniture construction has relied for the most part upon wood-to-wood joints. Some require nothing more than a wedging action to hold them fast. Several rely on wooden pins or wedges to help lock them in place, while most use some kind of adhesive.

Wood can be connected with staples, yet we're hesitant to call this joinery. What's the difference between cutting dovetail joints for a carcase or stapling it together? What makes one joinery and the other merely construction? And why use certain joints in one situation and not in another? The answers to these questions come from an understanding of the forces at work on the wood fibers.

To be successful, joints must accomplish several things, and they must do all of these simultaneously. They must provide mechanical resistance, offer good gluing surfaces, and allow for the expansion and contraction of wood in response to changes in humidity. That's a tall order, but it's exactly what good joinery accomplishes and what stapling does not.

Both of these joints provide mechanical resistance. The half-lap joint requires the use of fasteners or glue, while the dovetail can stay together on its own.

MECHANICAL RESISTANCE

Mechanical resistance can be achieved in various ways (see the photo above). Some joints can barely hold together on their own; a half-lap joint without fasteners or glue will fall apart. Others, such as a through dovetail, will hold together without even a wipe of glue. Yet when glued in place, both provide structural integrity.

There are several forces that a joint must be able to withstand in order to hold, and both the half-lap and the dovetail do this in their own fashion. These don't occur on a daily basis unless a piece is used often, like a chair or a kitchen table. But these stresses come heavily into action when a chair is tipped back, a table leg is kicked, or when something heavy is loaded into a cabinet.

COMPRESSION, TENSION, AND SHEAR

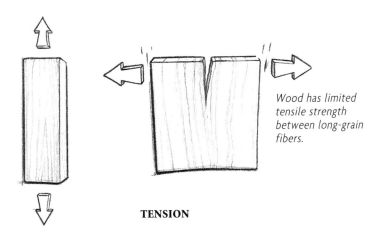

Compression on end grain has little effect.

Long-grain compression failure is common.

COMPRESSION

Wood has limited tensile strength between long-grain fibers.

TENSION

Shear occurs in two directions simultaneously.

SHEAR

Compression, tension, and shear are the main forces an object must endure (see the drawing at left). Compression force tries to compress or crush an object in on itself. Tension force tries to pull an object apart. Shear forces come into play when parts of an object are being moved in opposite directions. The force we call racking is actually a combination of these basic stresses.

Compression

There are several types of compression forces. One of these is compression parallel to the long grain, or simply force down onto the end grain of a board. Compression on the end grain of a board isn't much of a problem. Most furniture can handle this force before shattering under a weight. Unless your furniture doubles as a car jack, there is little cause for concern. According to R. Bruce Hoadley in his book, *Understanding Wood* (The Taunton Press, 1980), four $1/8$-in. hickory dowels can support a load of 250 lb. if no other forces are present.

Compression on the face or edge of a board, such as a hammer blow, can be more damaging (see the photo at left on the facing page). As we all know, wood dents when you drop a tool onto its surface. The same type of deformation can occur when a joint is put under racking pressures.

Tension

Wood is not nearly as strong in tension as it is in compression. This is why a chisel can split a board lengthwise but not across the grain.

Wood is strong in compression on the end of a board but not on its face, where dropping a maul leaves a very big impression.

Wood has limited tensile strength along the grain—a wedge tapped into a board will split it. And in a short-grain situation between two crosscuts, the wood between two crosscuts will split out easily along its long-grain fibers.

Examples abound. When you split firewood you take advantage of the wood's limited tensile strength. When you drive a wedge or nail into wood you can easily split the long-grain fibers apart (see the photo at top right). There is little strength between the long-grain fibers.

This is why short-grain situations, places where short runs of long grain are left unsupported, are problem areas and liable to break. My dad taught me about this once when we were putting up some wall studs. He was notching the sides of some 2x4s by making two crosscuts about halfway through and then knocking out the remaining chunk with his hammer. It worked like a charm because wood is so weak along its long-grain fibers.

When wood is joined in a way that allows for its shrinkage and expansion (see pp. 69-71), then tensile strength will not be an issue. When wood isn't joined

properly, the force on the wood may surpass its tensile strength, causing a split. It's not hard to find examples. Many antique stores have a piece or two where a wide solid panel of wood cracked because the wood was fastened or glued in place (see the photo above). You can minimize the effect of tension by designing joints within certain size limits and allowing large panels to move.

If a solid-wood panel is fastened to its frame, it will be unable to expand and contract in response to moisture changes in the environment. The wood may split at its weakest point. This panel was probably glued into grooves in the frame.

Shear

Shear occurs when forces act upon a surface from opposing directions (see the top drawing on the facing page). If pressure is applied down on a half-lap joint, the glue joint will have force on it. But at the same time, the joint will be trying to move up and away from the shoulder. The quality of the glue joint and the fit of its mating surfaces will determine how well it will respond to this strain.

Short-grain situations can also set up shear stresses. A dowel or fastener placed too close to the end grain of a board can shear out the short-grain fibers if stressed.

Racking

Racking forces are a common threat to furniture. Imagine your heaviest friend on your most delicate wooden chair. Now in your mind's eye tip your friend back on those two slender rear chair legs (see the bottom drawing on the facing page) and think about the forces that are being exerted on the chair-rail joints. The joints are trying to hold up the front of the chair like a cantilever while all that weight is levering down on the joint and compressing the long-grain fibers there. Better tip your friend back before it's too late.

When a delicate chair is rocked back onto its rear legs repeatedly, there is not only compression and tension acting on the joint but also shear. The glue is being asked to hold the joint intact as all these forces are being applied. As the front of the chair is cantilevered, the rear glue joints have to hold it up and keep the joint intact.

So how does a chair survive the repeated assaults of your large friend? If the chair is doweled together, chances are very good that it won't. The racking strain will cause the dowels' long-grain fibers to be compressed inside their holes. As the dowels start to deform and moisture cycling occurs, the glue joint will get pushed beyond its shear-strength limits. The dowels will loosen, your chair will fall apart, and your friend will not be smiling as he sprawls amidst the rubble of your chair. A tenoned joint would have been a better choice.

Guidelines for joinery

There are several ways you can engineer your joinery so it can more successfully resist the forces of compression, tension, and shear. Some joints, such as dovetails and wedged joints, offer a natural mechanical resistance; others rely on maximizing long-grain gluing surfaces. The through dovetail, with its angled tails and pins and long-grain gluing surfaces, offers the best possible alternative for a corner joint with strength. But the simple rabbet joint can also resist movement when cut and glued properly.

For the best resistance to mechanical forces, all joints need to fit well, and the mating surfaces need to be flat and clean. There should be no burn marks, loose fibers, tearout on the surface, or snipe at the ends of boards. Machining marks should be kept to a minimum.

A bewildering array of chemicals faces the woodworker ready to glue wood. For most general

SHEAR FAILURE

Downward pressure at one end of a half-lap will produce shearing forces on the glue joint.

Shear failure can also occur in short-grain situations when a fastener is placed too close to the end of a board.

RACKING

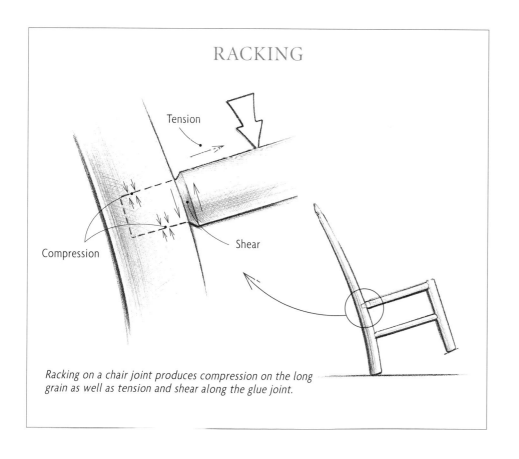

Tension

Compression

Shear

Racking on a chair joint produces compression on the long grain as well as tension and shear along the glue joint.

GLUING SURFACES

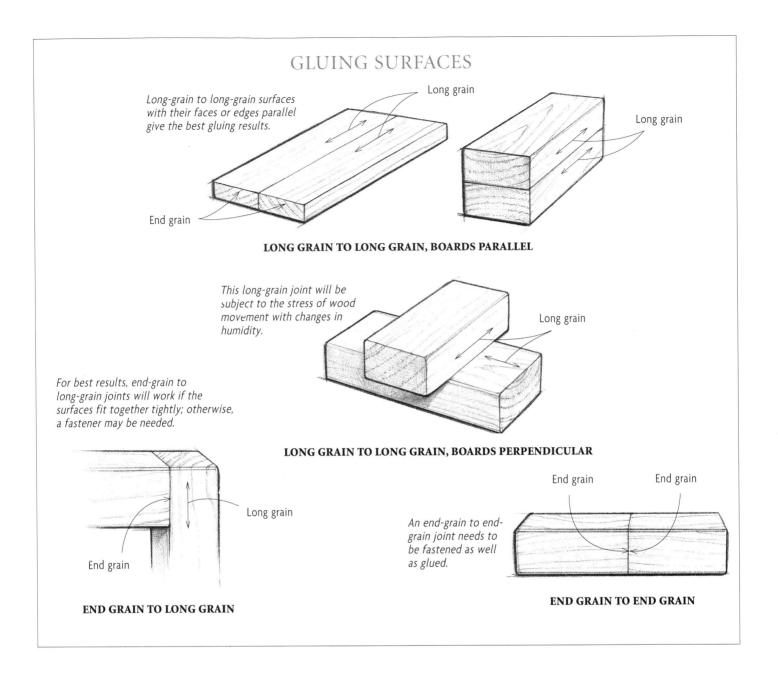

Long-grain to long-grain surfaces with their faces or edges parallel give the best gluing results.

Long grain

Long grain

End grain

LONG GRAIN TO LONG GRAIN, BOARDS PARALLEL

This long-grain joint will be subject to the stress of wood movement with changes in humidity.

Long grain

LONG GRAIN TO LONG GRAIN, BOARDS PERPENDICULAR

For best results, end-grain to long-grain joints will work if the surfaces fit together tightly; otherwise, a fastener may be needed.

Long grain

End grain

END GRAIN TO LONG GRAIN

End grain

End grain

An end-grain to end-grain joint needs to be fastened as well as glued.

END GRAIN TO END GRAIN

applications, I use a yellow PVA (aliphatic-resin) glue. Yellow glue is not gap filling and so demands close-fitting tolerances between surfaces to hold successfully. But when properly spread and cured, it creates a bond that is stronger than the wood itself. Yellow glue allows for a certain amount of creep within a joint. It's water soluble and can work in a fairly wide temperature range. But mostly it's strong and simple to use.

Whatever glue you choose, it needs to be fresh. I always buy my yellow glue in gallon containers and look for the newest label in the bunch of bottles or a manufacturing date if I can locate one. After a year, if the glue hasn't been used, I toss it out. Any yellow glue that has a stringy look to it is suspect and should be tossed. Always glue up a test piece if you're concerned about results.

A glue joint requires smooth yet porous wood surfaces that can readily accept the glue to form a bond. Long-grain to long-grain glue joints are optimal (see the drawing on the facing page). For the best results, they should be glued together with their edges parallel to one another. If they are glued at right angles, the joint will be good, but not quite as strong. Deeper joints that increase this gluing surface will help to increase strength (see the drawing at right). So will wider joints, up to the point where shrinkage becomes a concern. Increasing the number of joints by using more rails or members will help as well. Carcase joints such as dovetails or finger joints can be strengthened by using more of them per corner.

End-grain to long-grain glue joints are not optimal (the end grain tends to suck up moisture in the glue before the glue can do its holding job), but will work if the gluing surfaces fit tightly together. In a table apron, well-fitted shoulders on the table rails will help resist racking. In a rabbet or butt joint, a tight fit and fasteners really help the strength of the joint. Glued-in splines or keys that offer long-grain gluing surfaces can also help to increase strength (see pp. 108-111). Also be aware of the amount of wood that is removed to make a joint. The remaining wood must be strong enough to withstand the fitting of the joint and any subsequent racking of the piece.

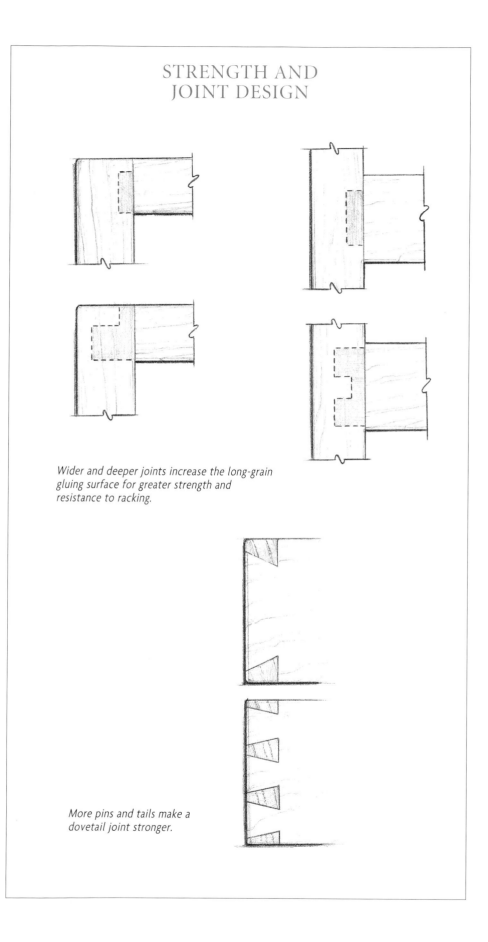

STRENGTH AND JOINT DESIGN

Wider and deeper joints increase the long-grain gluing surface for greater strength and resistance to racking.

More pins and tails make a dovetail joint stronger.

DIAGONAL RESISTANCE

A cabinet without a back has little resistance to racking.

Tight-fitting shoulders and deep, wide joints help prevent racking.

Wide front and back rails and a back stabilize the front and rear of this cabinet by triangulating.

Provide for diagonal resistance by triangulating with members, which also helps increase strength. A heavy wardrobe laden with objects is under a great deal of diagonal force when it's dragged across a floor. The wide expanse needed to provide room for the doors means there's little resistance to racking across the front of the cabinet (see the drawing at left). But by putting wide joints on the connecting rails of the piece and a stout back on the wardrobe, much of this diagonal racking can be successfully resisted.

Joint surfaces should always fit snugly without being so tight that there's no room for glue. A joint that fits well will require some pressure to be put together and some persuasion to be taken apart. Joint surfaces should also be freshly cut or dry-clamped together away from dirt and oxidation before gluing. Do not sand a joint to clean it unless you are very careful not to round it over and ruin the fit. Blow the dust away too before gluing with fresh glue.

Avoid the extremes when fitting a joint. Don't use a hammer to put joints together—it can split the surrounding wood. And don't force a joint home with clamps that doesn't want to fit. Try to make sure the joint goes together well when dry-clamped and before putting on glue. Overclamping when gluing can pull a joint out of alignment.

End-grain to end-grain joints can be used only if there is some kind of fastener or wedge providing the strength. Sizing the joint with a thin coating of glue to prevent too much glue absorption is another hedge against joint failure.

WOOD MOVEMENT: WARPING AND DISTORTION

As it dries, a board can distort in various ways.

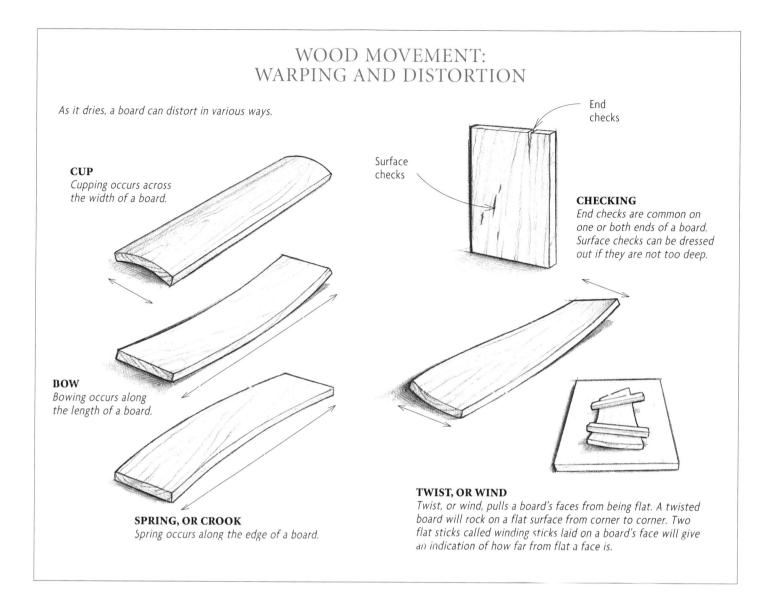

CUP
Cupping occurs across the width of a board.

BOW
Bowing occurs along the length of a board.

SPRING, OR CROOK
Spring occurs along the edge of a board.

CHECKING
End checks are common on one or both ends of a board. Surface checks can be dressed out if they are not too deep.

End checks

Surface checks

TWIST, OR WIND
Twist, or wind, pulls a board's faces from being flat. A twisted board will rock on a flat surface from corner to corner. Two flat sticks called winding sticks laid on a board's face will give an indication of how far from flat a face is.

EXPANSION AND CONTRACTION OF WOOD

Wood continues to respond to its environment long after it has been sawn, dried, milled, joined, sanded, and polished. This is one of wood's charms, but is also a source of frustration for the woodworker. Its hygroscopic nature means that it responds like a sponge to changes in humidity: It expands in humid weather and shrinks when it's dry. It cracks if it must, and twists and warps if stressed enough by its growth or the drying process (see the drawing above). And even

SHRINKAGE DIRECTIONS

Longitudinal shrinkage is negligible. Shrinkage across the growth rings (tangential shrinkage) is generally twice as much as between the growth rings (radial shrinkage).

Longitudinal

Tangential

Longitudinal

Radial

Radial

FLATSAWN BOARD

QUARTERSAWN BOARD

APPROXIMATE SHRINKAGE VALUES OF COMMON DOMESTIC HARDWOODS[1]

	TANGENTIAL[2] (%)	RADIAL[2] (%)	12-IN. SAMPLE[3] (IN.)
Alder, Red	7.3	4.4	.250
Ash, White	7.8	4.9	.267
Aspen, Quaking	6.7	3.5	.230
Basswood	9.3	6.6	.319
Beech	11.9	5.5	.408
Birch, Paper	8.6	6.3	.295
Birch, Yellow	9.5	7.3	.326
Buckeye, Yellow	8.1	3.6	.278
Butternut	6.4	3.4	.219
Cherry, Black	7.1	3.7	.243
Chestnut	6.7	3.4	.230
Cottonwood, Eastern	9.2	3.9	.315
Elm, American	7.2	4.2	.247
Elm, Rock	8.1	4.8	.278
Hackberry	8.9	4.8	.305
Hickory, Pecan	8.9	4.9	.305
Hickory, Shagbark	10.5	7.0	.360
Holly	9.9	4.8	.339
Honeylocust	6.6	4.2	.226
Locust, Black	7.2	4.6	.247
Madrone, Pacific	12.4	5.6	.425
Magnolia, Southern	6.6	5.4	.226
Maple, Bigleaf	7.1	3.7	.243
Maple, Red	8.2	4.0	.281
Maple, Sugar	9.9	4.8	.339
Oak, Northern Red	8.6	4.0	.295
Oak, Southern Red	11.3	4.7	.387
Oak, White	10.5	5.6	.360
Persimmon	11.2	7.9	.384
Sassafras	6.2	4.0	.213
Sweetgum	10.2	5.3	.350
Sycamore	8.4	5.0	.288
Tupelo, Black	8.7	5.1	.298
Tupelo, Water	7.6	4.2	.260
Walnut, Black	7.8	5.5	.267
Willow, Black	8.7	3.3	.298
Yellow Poplar	8.2	4.6	.281

[1]Adapted from Wood Handbook (U.S. Forest Service), and Understanding Wood (The Taunton Press, 1980)

[2]Approximate percent of shrinkage as wood moves from green to oven-dry moisture content

[3]Shrinkage in inches of a 12-in.-wide flatsawn board as it dries from 12% moisture content to 4%

though you're clever enough to finish off your newly made masterpiece with five coats of finish, this only slows the process down. It in no way prevents moisture loss or gain.

Wood expands and contracts more in certain directions than in others because of its cellular structure (see the bottom drawing on p. 65). Each year a new growth ring of wood cells is added onto the tree. The fibers in these cells are arranged like a group of straws bound together and filled with sap. The fiber ends are open and porous, soaking up and losing moisture as soon as they're cut into logs or boards. End grain is the name we give these porous fiber ends in the board; long grain is what we call them in their lengths. Not surprisingly, end grain absorbs and loses 12 times as much moisture as long grain.

Shrinkage in a piece of wood can occur in different directions and at different rates as the board soaks up or loses moisture. Shrinkage along the length of a board is called longitudinal shrinkage. This movement is so small (less than .1%) that woodworkers can ignore it. However, there can be considerable movement across the long-grain fibers. Shrinkage across the growth rings is called tangential shrinkage. Shrinkage between these rings is called radial shrinkage.

Species and the movement of wood

The charts on the facing page and at right gives shrinkage values for common domestic hardwoods and softwoods, respectively. As you can see, shrinkage varies considerably from species to species, from very stable woods, such as redwood and cedar, to woods that move a lot, such as madrone and beech. Notice also in the charts that tangential shrinkage and radial shrinkage differ within a species—generally tangential shrinkage is about twice as much as radial shrinkage.

The last column of each chart gives the shrinkage (in inches) you can expect in a 12-in.-wide flatsawn board that goes from a moisture content of 12% to 4%. This is a fairly large swing, akin to storing your wood out in the garage on a concrete floor and then bringing it into the house and setting it near the woodstove. But it gives an idea of how much each species can move in an extreme situation. The method used to figure out this potential for movement using shrinkage tables and standard formulas can be found in R. Bruce Hoadley's book *Understanding Wood*.

If charts and formulas make your head spin, there's a simpler (although slower) way to predict movement: Just watch what happens to some sample boards in your shop. Take boards of the species you plan to use in your project, cut them to width, put your standard finish on them, note their size, and watch them change in dimension throughout the year. This will give you some idea of how this species reacts in your part of the country.

APPROXIMATE SHRINKAGE VALUES OF COMMON DOMESTIC SOFTWOODS[1]

	TANGENTIAL[2] (%)	RADIAL[2] (%)	12-IN. SAMPLE[3] (IN.)
Baldcypress	6.2	3.8	.213
Cedar, Alaska	6.0	2.8	.206
Cedar, Eastern Red	4.7	3.1	.161
Cedar, Incense	5.2	3.3	.178
Cedar, Northern White	4.9	2.2	.168
Cedar, Port Orford	6.9	4.6	.237
Cedar, Western Red	5.0	2.4	.171
Douglas Fir, Western	7.5	4.8	.257
Fir, Balsam	6.9	2.9	.237
Fir, White	7.0	3.3	.240
Hemlock, Eastern	6.8	3.0	.233
Hemlock, Western	7.8	4.2	.267
Larch, Western	9.1	4.5	.312
Pine, Eastern White	6.1	2.1	.209
Pine, Pitch	7.1	4.0	.243
Pine, Ponderosa	6.2	3.9	.213
Pine, Sugar	5.6	2.9	.192
Pine, Western White	7.4	4.1	.254
Redwood	4.9	2.2	.168
Spruce, Red	7.8	3.8	.267
Spruce, Sitka	7.5	4.3	.257
Tamarack	7.4	3.7	.254

[1] Adapted from *Wood Handbook* (U.S. Forest Service), and *Understanding Wood* (The Taunton Press, 1980)

[2] Approximate percent of shrinkage as wood moves from green to oven-dry moisture content

[3] Shrinkage in inches of a 12-in.-wide flatsawn board as it dries from 12% moisture content to 4%

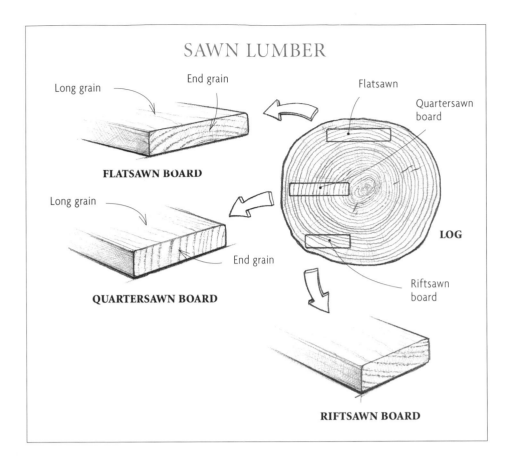

SAWN LUMBER

Long grain

End grain

FLATSAWN BOARD

Long grain

End grain

QUARTERSAWN BOARD

Flatsawn

Quartersawn board

LOG

Riftsawn board

RIFTSAWN BOARD

How a board is sawn

Boards can be sawn from a log in several different ways (see the drawing above). In flatsawn, or plainsawn, lumber, the growth rings show across the end grain in a mostly horizontal pattern. Since wood shrinks more across its growth rings, flatsawn boards are more susceptible to movement and cupping across their widths. Flatsawn lumber generally has a pronounced grain pattern that appears flame-like or elliptical.

In quartersawn and riftsawn lumber, the growth rings are arranged anywhere from vertical to about a 45° angle to the board face. With these methods of sawing, fewer boards can be cut out of a tree, so quartersawn and riftsawn boards are rarer and more expensive than

flatsawn boards. But they yield a material that shrinks much less across its width because of the arrangement of its growth rings.

Quartersawn lumber yields a regular straight-grain pattern. Riftsawn lumber, which is cut closer to a 45° angle, is prized for its ray-fleck figure, especially in white oak, beech, and sycamore.

Guidelines for joinery

Long ago, I worked in a shop where another woodworker had dovetailed together a marvelous cabinet in white oak. It could act as a room divider because of its finely finished solid white oak back. Unfortunately he had glued on this back. So when the piece got to its new home and the back expanded, it cracked the cabinet apart.

Gluing on solid backs to carcases or bottoms to boxes or chests is asking for trouble. Laminations in the back itself can split or the piece can delaminate from the carcase. If the glue joints hold, then the box itself can rack or warp so that lids or doors no longer close properly. Or it can just blow the whole carcase apart. The larger the pieces of wood involved, of course, the greater the potential for disaster.

Ignoring wood movement when you design a piece of furniture guarantees any number of possible problems: joint failure, improperly fitting doors and drawers, racking, or split panels. How can joinery prevent any of this?

In a carcase piece, grain orientation is critical. Arrange boards so their grain will shrink and expand as a unit, not struggle to pull the carcase apart (see the drawing below). One guideline for orienting grain direction in all kinds of furniture is to let length determine grain; that is, the grain direction should follow the longest dimension of a piece. Now this may not be the wisest choice for all the pieces of a carcase, for instance, the short ends of a deep box. But the advice is certainly worth following for frame-and-panel work.

In frame or stool construction, the members are smaller than in carcase construction, so shrinkage and expansion are less problematic. If you orient the members properly, you minimize problems even further. Frames that hold panels must allow for the movement of the panels. Panels that are allowed to float in a groove will not pull themselves apart

Be careful of mixing materials as well. Using different species of wood in a project can create problems if their shrinkage rates are dissimilar. Mixing plainsawn and quartersawn wood may also cause problems because of their different movement values. Similarly, combining solid stock and plywood or other composite materials needs careful planning. Composite materials don't shrink or expand much; when used in conjunction with solid stock, they must not prevent the movement of the solid wood.

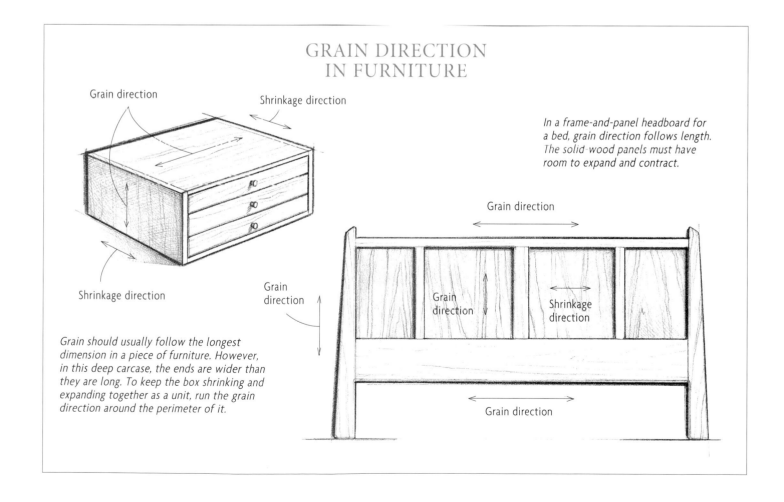

GRAIN DIRECTION IN FURNITURE

Grain direction

Shrinkage direction

Shrinkage direction

Grain direction

Grain should usually follow the longest dimension in a piece of furniture. However, in this deep carcase, the ends are wider than they are long. To keep the box shrinking and expanding together as a unit, run the grain direction around the perimeter of it.

In a frame-and-panel headboard for a bed, grain direction follows length. The solid-wood panels must have room to expand and contract.

Grain direction

Grain direction

Shrinkage direction

Grain direction

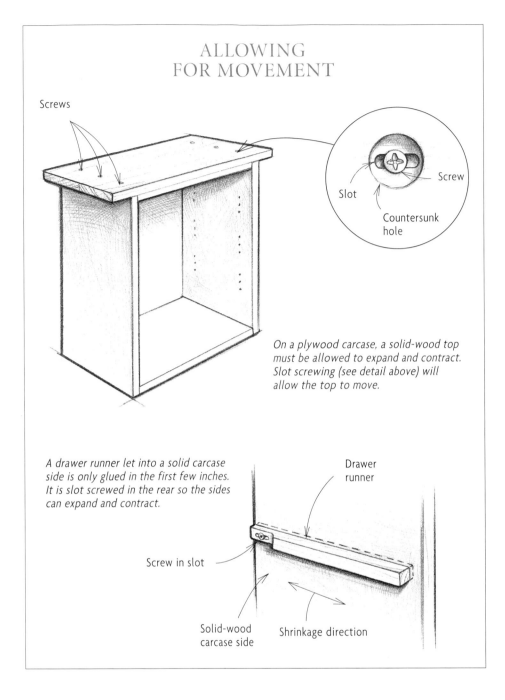

ALLOWING FOR MOVEMENT

Screws

Slot

Screw

Countersunk hole

On a plywood carcase, a solid-wood top must be allowed to expand and contract. Slot screwing (see detail above) will allow the top to move.

A drawer runner let into a solid carcase side is only glued in the first few inches. It is slot screwed in the rear so the sides can expand and contract.

Drawer runner

Screw in slot

Solid-wood carcase side

Shrinkage direction

holds true for a solid-wood cabinet side onto which a drawer framework is added. The drawer runners fit into grooves cut into the carcase sides, but they cannot be glued entirely into place. Only the first few inches or so are glued and the back of the frame is slot-screwed into the carcase, thereby allowing for movement (see the drawing at left).

Don't use wide tenons on rails or aprons that are especially large, since wide tenons can move too much in a mortise, causing splitting or glue failures. Instead, break them up into several smaller tenons that will move independently. Limit the width of tenons to 3 in.

Always give solid-wood doors, panels, and drawer sides room to expand if built in a dry season. And pre-finish panels before setting them into their frames so any shrinkage won't reveal an unsightly finish line if the panel pulls out of the frame.

Because end grain soaks up and loses more moisture than long grain, many tabletops have breadboard ends covering their porous end-grain surfaces. These are left unglued to permit movement, but they slow down this movement and help to keep the tops flat. Through joints showing end grain are strengthened and seem to survive moisture changes better with the addition of wedges.

Wood moves—it's probably one of the few things all woodworkers agree upon. The best way to deal with this movement is to plan for it in your designs and allow for it in your joinery.

Always allow for movement across wide boards. Anything that gets added onto them must allow for movement by either being slotted or slot-screwed on and not glued. This is especially important when placing a solid-wood tabletop onto a leg-and-apron framework, or a solid top onto a plywood case. The top must be allowed to move throughout the seasons. The same

CHAPTER SIX

JOINERY IN FURNITURE CONSTRUCTION

Joint selection depends on many factors, including function, skill, and design. This Tansu chest by the author could have been constructed in several different ways. Photo by Phil Harris.

One of the difficult things about building furniture can be comprehending all the parts and pieces that go into the making. Where should everything fit, and what's the best way to put all those parts together? There are always several ways of building a table, but what's the best one with so many joints and construction methods to choose from? Furniture is built with a strategy for construction, and the maker must plan for its assembly.

CHOOSING THE RIGHT JOINT

Joint selection depends on several factors: function, economy, skill, and design. What will the piece do, and how fast does it need to be completed? What does the woodworker know or have to cut, and what effect will the joinery have on the design of the piece?

Joinery plays many roles besides providing a structural framework upon which a piece of furniture hangs. It gives strength to pieces, allows for wood movement, helps with assembly, and influences design. These factors make it crucial for a woodworker to understand. And the more joints that are at your disposal, the greater the variety of pieces you can build.

Function

Every violin is not a Stradivarius. Each maker may of course strive to that level, but some just make fiddles. And they're happy to be doing so. A piece of furniture can make beautiful music, too, or just sit quietly in the background. Some pieces are quickly built to be just good enough for the job. Others want that attention to detail, the heirloom touch, that a concerned maker's hand can bring to a beautiful piece.

So the question of function is really two questions: What will the piece of furniture do, and how will it get the job done? Some pieces that hold together for ten years are considered a great success. Others are designed and constructed to last through the changes of styles and generations. A maker will always have to decide where a particular piece stands on this issue.

The purpose of the piece will determine how strong a joint is needed. If the piece is a picture frame, then slot dovetails aren't really required for strength at the corners. Miters will do just fine. On the other hand, a tool chest or dresser drawer might need the real strength a dovetail joint offers. Ask yourself how the piece needs to function, how long it needs to keep standing, and what you want from the building of it. All of these answers speak to the function of a piece.

Economy

"How quickly does the joint need to be cut?" asks the question of economy. Is time critical, as it is for the professional, or is the joy in the making, no matter the cost in time? One joinery option can take an hour and another days, so this is a question that needs a personal answer. The purpose of some pieces is to provide a canvas for marquetry, carving, or a type of finish. These pieces need care in their making, but the joinery is not the focus of the piece. Joints are cut carefully but economically in terms of time.

Other pieces are made for the sheer enjoyment of the building. The more complicated they are, the better, as they provide an excuse to spend more time in the shop. Some woodworkers take great pride in showing off their hard work, keen eye, and precise joinery. The time required reflects the effort.

Skill

Familiarity breeds contempt, they say. But familiarity with a tool often brings far more contentment than contempt. One may not always be looking for a challenge when building a piece. Going with what you know can move a job along more quickly and for that reason be more satisfying. The skill a woodworker brings to the project will then influence choices in joinery.

Consider this. The body has a memory as surely as the mind. The first time through any operation you discover where to place your hands and set your feet. The times after that it's easier and quicker to do. The day comes when you don't even notice what you're doing anymore. Your body knows how to stand; your hands move automatically to the proper place on the tool. As more techniques become familiar, they get added to the vocabulary of building. This is what sets apart the master, who has the range of skills to choose the method appropriate for the job and the setting.

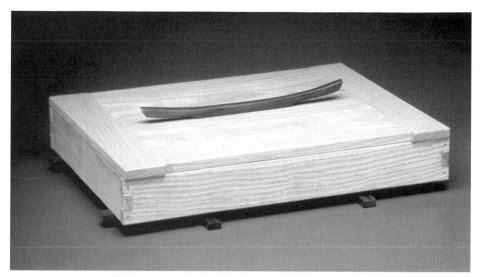

This small jewelry box, designed and made by the author, uses the same carcase joinery that could be used on a much larger cabinet. Photo by Harold Wood.

As for the tools available, these can limit options even more dramatically. Perhaps your dovetail saw is dull, but your router bit is brand new. It's an easy choice here. If a maker is capable only of cutting dadoes, then the work will reflect that. I know that when I first started out, my first tool was a radial-arm saw. I built a lot of stuff with dadoes and rabbets. But as skill and knowledge increase, so does the satisfaction in your work, and potentially the scope of the designs.

Design

An overlooked aspect of joinery is how it helps to influence design. The most obvious example is the choice one makes when building something as basic as a drawer. I can give you a dozen ways of doing the job, and none would be the wrong choice. But each would influence how you view the piece, and how you approach and handle

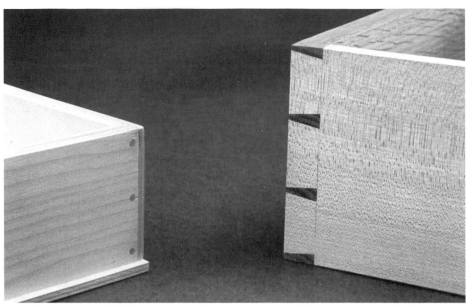

The shop drawer made with pinned rabbet joints (left) and the dovetailed dresser drawer (right) show a difference in how they were made and how they will be seen, handled, and used.

Small details, such as the setbacks and reveals on the Tansu chest (see the photo on p. 71), add interest to a piece. Photo by Phil Harris.

it. Some joints, like a nailed rabbet joint, shout speed and economy, while a half-blind dovetail is the epitome of restraint and precise skill.

Details as simple as reveals or setbacks can produce shadow lines and add interest in a piece. This is a detail that has to be planned for in the joinery. Flush surfaces in a joint, even when well sanded, rarely stay flush as wood moves.

Another useful feature of joinery is how it helps in assembling furniture. A rabbet joint gives even the worst builder an advantage over someone using a simple butt joint. The shoulders of the joints help to line up the boards so that gluing or fastening is made easier. In more complex situations, joints can help by squaring carcases or frames. For example, through dovetail joints will hold a carcase together so that subsequent measuring and assemblies can occur in or around it.

STRUCTURAL SYSTEMS

Furniture construction can seem complex and joints difficult to choose. But one way of understanding joinery is to understand the structural systems that form it. Being able to recognize these basic systems and the joints that can be used throughout them will make designing and completing your furniture projects far simpler.

There are many ways of categorizing furniture pieces. You could say that there are just cabinets, tables, and chairs. But there are end tables, side tables, and dining tables. There aren't just a few types of chairs but dozens, not to mention all the boxes, beds, and desks. But if you take a step back to get a broad overview of furniture construction, there seem to be only three basic types: carcase, frame and panel, and stool and table construction.

These three broad types of construction include most of the ways a piece of furniture can be put together. What becomes obvious as we look at these major furniture types is that there is overlap in how things can be built between categories. It turns out that many joints can be used in a variety of situations.

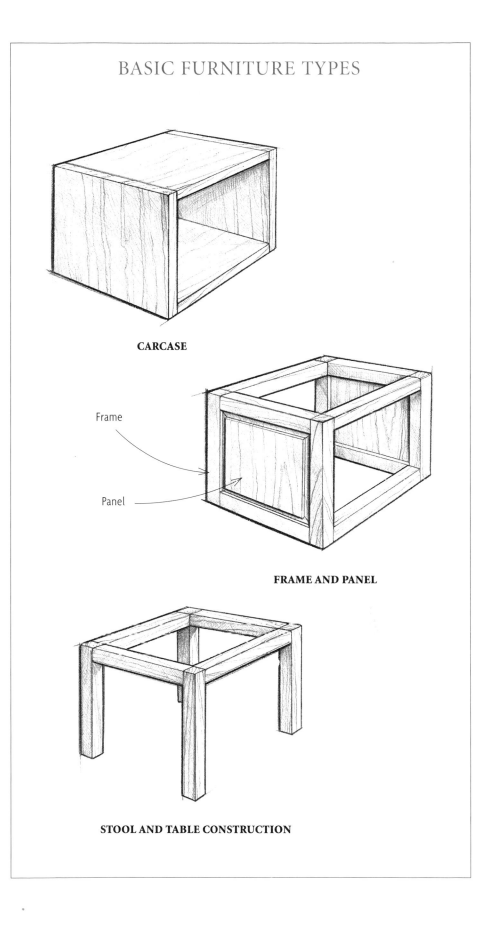

BASIC FURNITURE TYPES

CARCASE

Frame

Panel

FRAME AND PANEL

STOOL AND TABLE CONSTRUCTION

This collector's chest by the author uses panels joined at the corners to form a carcase. Drawers fill up the inside of this box. Photo by Harold Wood.

strength or movement. But if made of solid wood, grain direction becomes important. Since wood moves most across its width, a box must have all its sides oriented so that the grain direction runs around the perimeter of the box, as seen in the drawing on the facing page.

It may seem that the long-grain edges of these boards will allow for a really good joint. But over time, this nice square box will end up turning into a diamond shape as it tries to accommodate the wood movement. As the wood absorbs and loses moisture, the sides will be trying to push or pull themselves apart. A cabinet built this way will eventually have doors that won't close and drawers that won't fit.

Carcase joinery with the router is discussed in Chapter 7.

Carcase construction

Carcase construction is quite simply box construction. We use boxes to store our goods and to keep our treasures safe and hidden from view. These cabinets can be large and sit on the floor like dressers or chests, or they can rest on a table like a jewelry box. They can be filled with drawers, divided with shelves, or hung with doors. Some carcases are as small as a pipe rack, others are as large as an armoire. Cabinets can stand freely in a room, rest upon a table or frame, or hang on a wall. But the one thing they all share in common is the use of panels fastened together at their corners.

Carcases can be made of either solid wood or plywood or other composite material. When made of plywood or MDF, there are no concerns over grain direction and

Frame-and-panel construction

What envy fills woodworkers today when thoughts of using 20-in.-wide boards for a carcase cross their minds! Imagine having a seemingly endless supply of these wide beauties, like some woodworkers in the past. But even if there were such a supply of wood today, there are other issues that using wide stock for carcase sides brings up.

Those fortunate woodworkers of long ago had more problems with these wide boards than just figuring out where to store these treasures. Wide stock always moves more than narrow boards with fluctuations in humidity. It not only expands and contracts, but it can also cup and warp or crack and twist. And wide boards are heavy. Large carcase pieces could end up weighing several hundred pounds

CARCASE CONSTRUCTION
IN SOLID WOOD

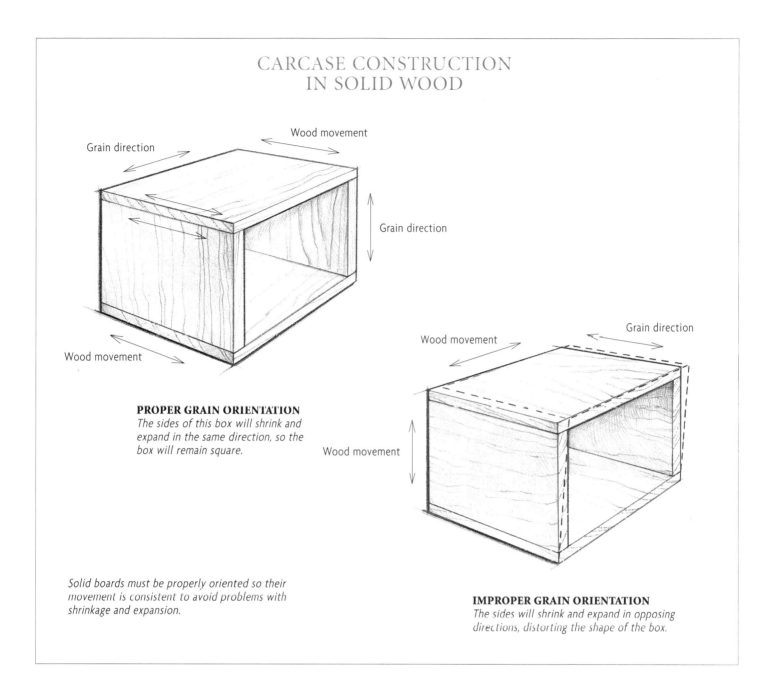

Grain direction

Wood movement

Grain direction

Wood movement

PROPER GRAIN ORIENTATION
The sides of this box will shrink and expand in the same direction, so the box will remain square.

Solid boards must be properly oriented so their movement is consistent to avoid problems with shrinkage and expansion.

Wood movement

Grain direction

Wood movement

IMPROPER GRAIN ORIENTATION
The sides will shrink and expand in opposing directions, distorting the shape of the box.

because solid boards needed to be thick enough for fasteners or joints to hold them together effectively.

Faced with these problems, some clever woodworker in the 15th century finally rediscovered what the Egyptians had known some 2,500 years earlier. This was that board-and-batten (solid wood)

construction could be refined not only in terms of engineering but also in terms of style. The visual weight as well as the physical weight of a piece could be lightened by the use of frames and panels. And a different type of structural integrity could be created.

Frame-and-panel work is a refinement both in style and weight over simple board-and-batten construction. This china cabinet by the author is built of frames connected together. Photo by Jim Piper.

would be able to give a solid look without the shrinkage and weight problems of wide boards, and panels could be flat or raised to provide even more visual interest. The frame edges could also be left plain or could be shaped, or have molding applied to them.

Another distinct advantage of this type of construction was that doors could be made up with frames and panels. Seasonal changes would no longer greatly affect their size, as had always been the case with wide solid-wood doors. Frame-and-panel work is used for building cabinets of course, but it is also used to construct tables, beds, chests, and mirror frames.

Frame-and-panel joinery with the router is discussed in Chapter 8.

Stool and table construction

We do one or two things with all of our furniture pieces. We set things in them or on them. Sometimes we do both. We put our goods or treasures inside of cabinets or drawers to keep them safe. Everything else that we build is some kind of platform. It sounds simplistic, but it's true. If it's not a box, then it's a platform on top of which we set ourselves or another object.

Seating pieces are platforms designed to raise our bodies up and off the floor. Whether we are seated for dinner or study, in casual or regal surroundings, chairs of infinite variety have been designed to lift us up, more or less in comfort.

The solution was to use smaller frame members to capture floating panels of solid wood (see the drawing on the facing page). The frame members would be solid wood of course, but so narrow in width that their shrinkage would be negligible. The panels could be put into grooves cut into these frame members and left unglued to allow for movement. This way they

FRAME-AND-PANEL CONSTRUCTION

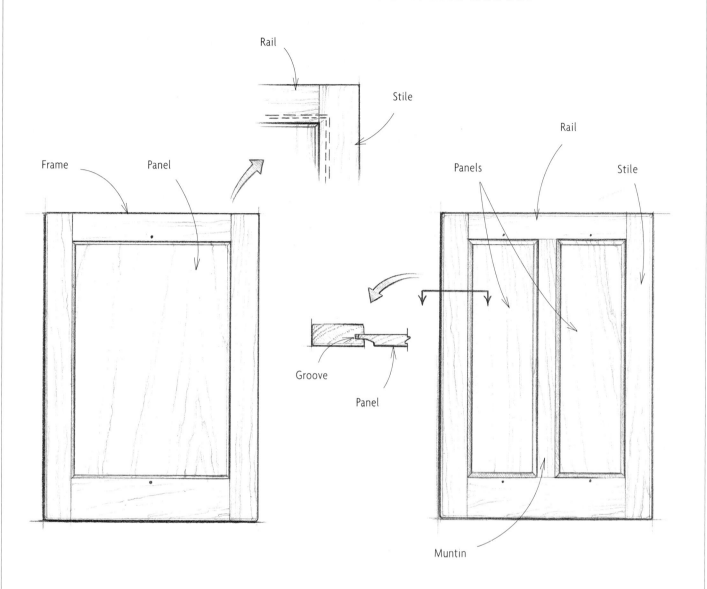

Rail

Stile

Frame

Panel

Rail

Stile

Panels

Groove

Panel

Muntin

Frame-and-panel work uses smaller members, called rails and stiles, to form the frames. Inside these frames, captured in a groove or let into a rabbet, sit the solid-wood panels, which are free to expand and contract.

Frames can be further divided by interior rails and stiles. (An interior stile is called a muntin.) Smaller panels shrink and expand less. Panels are often pinned in their center to prevent shifting or rattling.

Benches and tables act as platforms, too. They usually are capable of supporting some weight, which may turn out to be dinner for 12 or a set of carving tools. But all these structures share the common fact that they raise a surface up off the floor using legs.

There are two broad categories within this structural system: leg-and-rail construction and trestle construction (see the drawing on the facing page). Leg-and-rail (also called leg-and-apron) construction is similar to frame construction. In pieces like chairs, tables, and stools, frames are put together, but instead of doubling up posts at the corners, these frames share a common leg at their corners. Usually there are no panels inside these frames, and the bottom rail is generally omitted.

Platforms like these seating pieces can assume different postures and heights. The stool and the club chair were both built by the author.

This gate-leg table by the author uses eight legs in all to help hold up the top and its leaves. Collection of Todd Oppenheimer. Photo by Harold Wood.

STOOL AND TABLE CONSTRUCTION

Leg

Rail or apron

Built like frames, these forms share a common leg at their corners while often omitting the bottom rail or any panel.

LEG-AND-RAIL CONSTRUCTION

Stretcher or rail

Trestle or pedestal

A trestle bench is built like a carcase. A trestle table has a stretcher and two trestles or pedestals that support a tabletop.

TRESTLE CONSTRUCTION

In trestle construction, panels of varying sizes can be used to form seating pieces or tables that look similar to carcase work. They just eliminate that fourth side of the box. Pedestals can also be used to support table surfaces. For example, trestle tables have large flat tops resting on pedestals joined together by stretchers or rails.

Stool and table construction with the router is discussed in Chapter 9.

PLANNING THE JOINERY

One aspect of furniture making that is as simple as it is overlooked is planning. Nothing slows a maker down more in a shop than head scratching—not machinery breakdowns or power outages, not telephone calls or insistent spouses. And the results of not planning ahead can be disastrous.

Scale drawings and models

Drawings and models are important aids in planning a design. Two-dimensional representations can give a designer the information needed to correct or adjust a design that began as a doodle on a napkin. Whether $\frac{1}{8}$-in. scale, $\frac{1}{4}$-in. scale, or full scale, drawings of a design are useful expressions of what a maker hopes will occur in the wood. Various points of view can be drawn out, including side, front, and end elevations as well as plan details at the top and bottom of a piece (see the drawing on the facing page).

Once a shape or form is decided upon, you need to decide what kind of structure this form will require. As we've seen, there are only a few basic ways of building things. But a table, for instance, can take on so many different shapes that choosing a structure can really help in the design process. From this point you can then decide if you can actually build the piece. The simple question of how has put a quick stop to many a project. When the dreams of design meet the reality of the shop, some projects don't make the cut.

With the type of structure in place and a reality check taken, you can start sketching or drawing to scale to see how the parts and pieces add up to a whole. Often, drawings are only the beginning of the process. Models yield an enormous amount of feedback even when done to scale. And they require that you think about putting the piece together. This step-by-step procedure will make you think about the actual processes that will be needed later on to build the piece full scale.

A ¼-in.-scale model of a mahogany table points out the strengths and limitations of the design. It also helps to plan out the construction process.

SCALE DRAWINGS

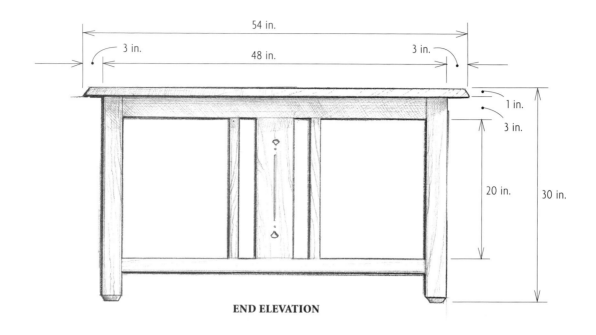

END ELEVATION

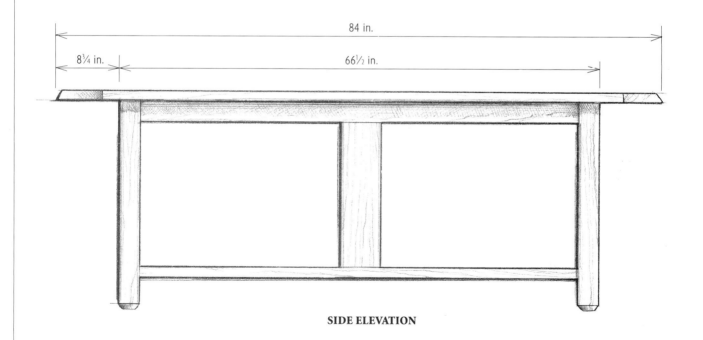

SIDE ELEVATION

Drawings of the library table shown in the photo on p. 56 helped to plan for its construction. Elevations done to scale give information useful in designing and building a piece.

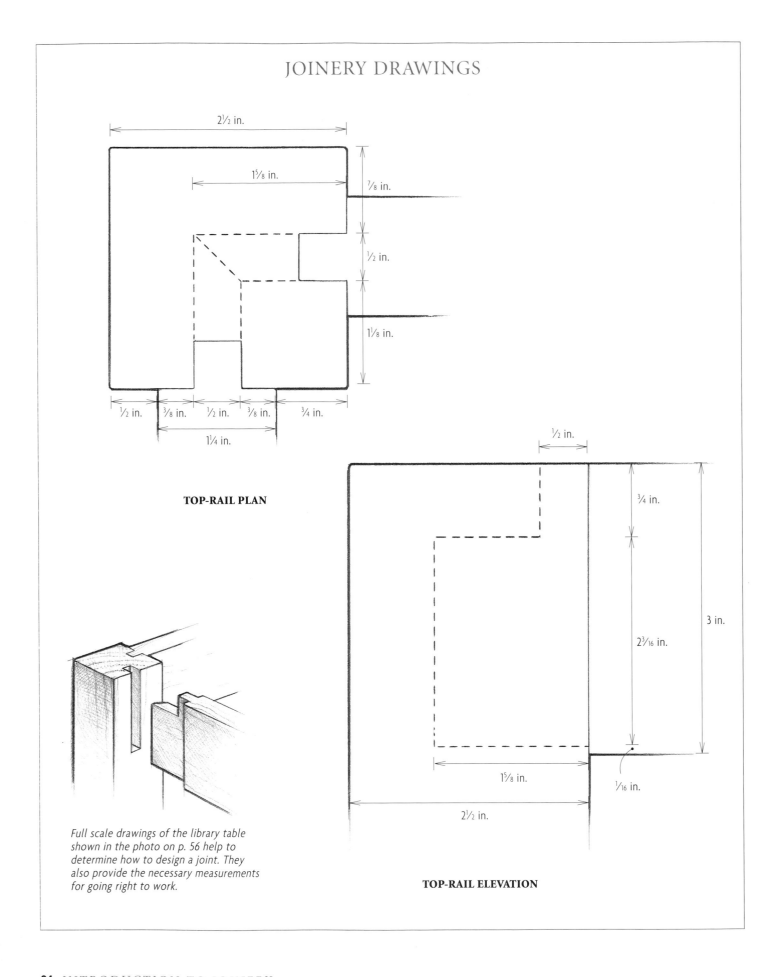

2½ in.

1⅝ in.

⅞ in.

½ in.

1⅛ in.

½ in. ⅜ in. ½ in. ⅜ in. ¾ in.

1¼ in.

TOP-RAIL PLAN

½ in.

¾ in.

3 in.

2³⁄₁₆ in.

1⅝ in.

1⁄₁₆ in.

2½ in.

Full scale drawings of the library table shown in the photo on p. 56 help to determine how to design a joint. They also provide the necessary measurements for going right to work.

TOP-RAIL ELEVATION

Full-scale joinery drawings and mockups

When I get to the stage in the design where I'm fairly certain of the shape and dimensions of a piece, I then start to consider how I'll put it together. Joints for many pieces often meet up in the same plane inside a leg, or a design will require that several functions occur right at a common location. I use full-scale joinery drawings for these situations to help me see the potential problem areas.

Full-scale drawings let you see how everything is fitting together, whether there's enough material to avoid a weak joint, and how the joinery will affect the overall dimensions of the pieces. Also, these drawings show exactly where the cuts will go. When it's time to cut the joints, you can

refer back to your drawing, take measurements straight from it and go right to work. The head scratching has already been done. The drawings I use to plan my joints generally take several views of the joint (see the drawing on the facing page). Front and side elevations show sizes, depths, and details. Plan views give joint widths and other information, such as tenon set-ins.

Mockups of joints are also useful for gathering information. If you're not certain how to cut a joint or how strong the joint might be, doing a full-scale mockup or model is a great way of finding these things out. When you actually see how much wood needs to be removed and how much is left, you can gain a much better sense of the strength of the joint.

CHAPTER SEVEN

CARCASE CONSTRUCTION

This oak chest of drawers (left) exhibits a more refined approach to construction than does this shop cabinet (right), but both are examples of carcase construction designed to fit a particular need.

There are plenty of ways to build a box. Every woodworker has banged together one sort or another, but cabinets and carcases seem to be a different and more complex matter. Carcase construction, however, simply involves choosing a method for putting a box together that is efficient, suits your design, and is as lasting as the amount of effort you want to put into it. The router methods that follow are the ones I use, starting with the simplest joints and moving to the more complex.

BUTT JOINTS

Put the end of one board up to the face of another with enough glue between these surfaces, and the result is actually a joint called the butt joint (see the drawing at right). It seems too simple to work, and over the long haul it is. But for a quick way of putting together a frame or box, the butt joint does the job. Because it joins long-grain and end-grain surfaces, glue and fasteners are required for the best results.

If the end-grain cut isn't square however, the biggest nail you've got won't help this joint. For a square box, the board's ends must be cut square both to the edge and to the face. While there are many saws that can make this cut, every once in a while I run into a large panel or a cumbersome piece that's easier to clean up with a router rather than with a saw. Here's a simple trimming method.

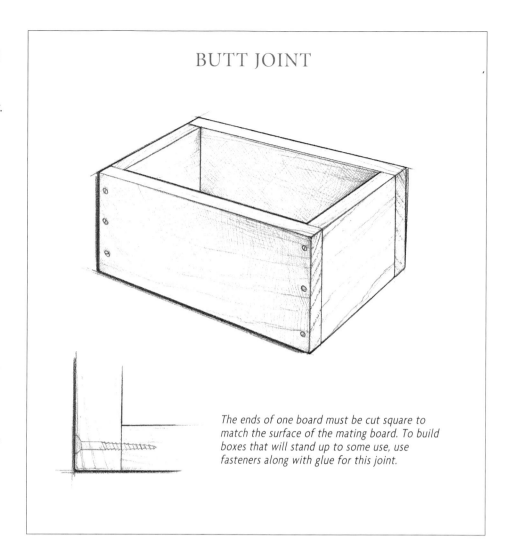

BUTT JOINT

The ends of one board must be cut square to match the surface of the mating board. To build boxes that will stand up to some use, use fasteners along with glue for this joint.

On a workpiece too wide for the table saw, end-grain cuts can be handled more easily with a router. The edge of this laminated tabletop can be routed square, if a straight piece of scrap is clamped underneath it to guide the flush-trimming bit.

End-grain cuts

The right-angle jig shown in the drawing on p. 44 can be used for an end cut if the jig is long enough. Start with a board with its long sides already milled parallel to one another, and square a pencil line across the board where you want your cut to go. If you're using a template guide, measure the offset between the edge of the straight bit and the outside of the guide. Set the right-angle jig that distance away from your pencil mark, and clamp it tight against the edge of the board or panel.

As always, a series of light passes is preferable to one big hogging pass on the end grain. Remember as well that end grain tends to burn with too slow a feed rate or a dull bit. Protect the end of the cut with a backer board to prevent tearout.

For boards that are too wide for the jig, use a long straightedge as a guide with a flush-trimming bit in the router (see the photo above left). The bearing on the bit will ride against the straightedge and guide the end cut. Since the cut is always flush to the bearing, any template shape, whether straight or curved, can be cut in this way.

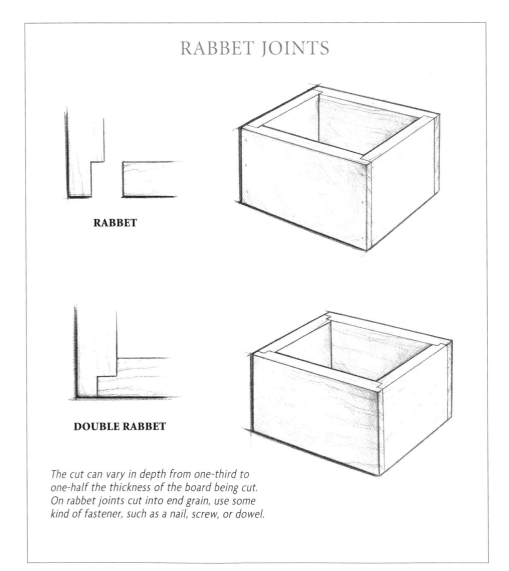

RABBET JOINTS

RABBET

DOUBLE RABBET

The cut can vary in depth from one-third to one-half the thickness of the board being cut. On rabbet joints cut into end grain, use some kind of fastener, such as a nail, screw, or dowel.

RABBETS

Butt joints are frustrating to put together because they slip around instead of lining up neatly for gluing or fastening. A simple shoulder cut or rabbet into the end or edge of one board helps to locate your corners and square up your box while increasing the overall gluing surface of the joint. A variation of this cut is the double rabbet, where rabbets are cut into the ends of both boards.

You cut the joint by routing with a straight bit down on the end or edge of a board. In solid wood, set the rabbet for slightly less than the thickness of the stock. This will allow the face surface to be slightly proud. But this makes it easier to pull the joint in tight with a clamp right over it. It's also easy to clean up that face grain later on. In plywood, with the risk of sanding or planing through a veneer, make the rabbet exactly the thickness of the stock or just a hair deeper.

Topside router cuts

The simplest way to cut the rabbet joint is with a bit as wide as the rabbet. Of course, any time the router is used topside, the workpiece must be securely fastened to the bench top, and it must be large enough to support the router on top of it. Mount a straight fence to the router base to index the cut from the end of the board. Add a wooden auxiliary fence to it to get better support and to provide a recess for the bit if necessary. Cut the recess by moving the wooden fence just under the bit and moving the bit slowly down into it.

If you are using a plunge router, set the depth of cut for the full depth you want, and set the fence for the thickness of the stock. If you've pencil-marked the joint, you can double-check these settings right down on the wood. The rabbet is

A rabbet cut made in plywood. To help prevent tearout, stop just short of the end, then pull the router back into the board to complete the cut.

cut on each end of a board being used in a box, but only two of the boards of a box get cut. Clearly mark which of these boards gets the cut.

Make a series of passes to get down to depth, moving the router from left to right across the board. To prevent tearout at the end of the cut, stop the pass before going off the end, and pull the router back into the work for that short distance.

RABBETED TOOLBOX

Rabbet joints are perfect for plywood construction, and I used them in the toolbox shown in the photo at right. I put together this toolbox to hold one of my routers. The box is made of ½-in. plywood with rabbeted corners. As shown in the drawing on the facing page, the top and bottom are grooved into the side and end pieces.

Start by rough-cutting the plywood parts about ¼ in. oversize in width and length. A full sheet of plywood is too tough to push past the blade at an exact final dimension. Rough out the pieces first, and then mill exactly to length and width.

This toolbox for a router is made of ½-in. plywood rabbeted together. The joints have been reinforced with ⅛-in. dowels in anticipation of heavy use.

To rout the rabbets, put a ½-in. straight bit into a plunge router, but check the thickness of the stock to see if it's over or under ½ in. thick. Set your straight fence to make a cut that is exactly the thickness of the stock. Set the depth of cut for half the thickness of the stock.

Mark out the face sides of all the pieces and indicate the two pieces that will get the rabbet cuts. Clamp one of them down to a bench top, making sure the clamp heads or bench dogs won't get in the way of the router. Then rout the rabbets from left to right on both pieces, making several passes to get down to depth.

The grooves for the top and bottom are more easily routed on the router table. Since my plywood was not quite a full ½ in., I mounted a ⅜-in. bit. Instead of using a spacer against the fence, I used another method. I took two passes to get a groove wide enough to fit the plywood. I cut the first groove on all the

pieces, reset the fence distance about ⅛ in. over, and finish-cut all the grooves to fit the plywood.

The depth of cut on the grooves is no deeper than the rabbet cuts. This means that all the grooves can go right through each piece without showing on the ends.

Another method for cutting small topside rabbets uses the bit made expressly for the job. The bearing on a rabbeting bit will preset the width of a rabbet, however, so it must match the thickness of the stock to be most effective.

Router-table cuts with a fence
Narrow pieces can be rabbeted cross grain more easily on the router table than topside with a hand-held router. A router-table setup can make the rabbet cut with either the rabbeting bit or a regular straight bit used with a fence. Several boards ganged together and held tight up against the fence

When all the rabbets and grooves are routed, put the box together to double-check the size of the top and bottom, and then cut them to fit in their grooves. If you need to, you can always trim a little off them to get them to fit a groove that is too shallow. Pull out all the clamps you'll need to glue up, and use wood cauls to spread out the clamping pressure.

The box is glued and then cut apart on the table saw to yield a framed-in top and bottom. That's why there's that extra ⅛ in. built into the height of the box, for the thickness of the saw cut. Set the blade height just under the thickness of the plywood so the top doesn't bind on the blade on that final cut. Use a knife to cut through the last fibers to separate the sections. If you want, the rabbets can be strengthened with dowels, nails, or screws put in at an angle. After assembly, make a stopped cut on the bottom of the box to create "feet."

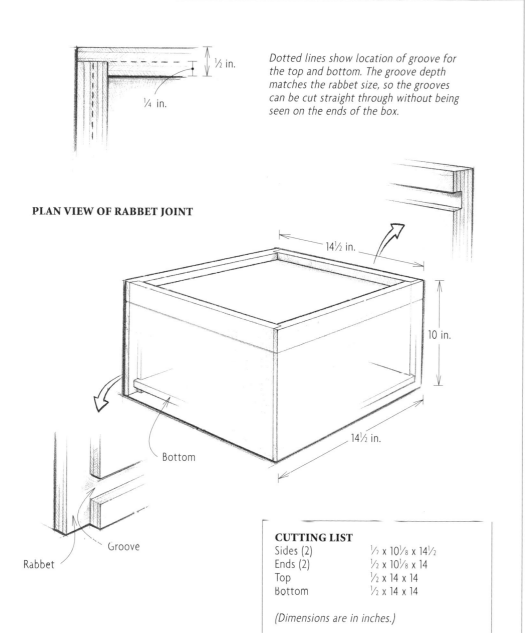

PLAN VIEW OF RABBET JOINT

½ in.

¼ in.

Dotted lines show location of groove for the top and bottom. The groove depth matches the rabbet size, so the grooves can be cut straight through without being seen on the ends of the box.

14½ in.

10 in.

14½ in.

Bottom

Groove

Rabbet

CUTTING LIST

Sides (2)	½ x 10⅛ x 14½
Ends (2)	½ x 10⅛ x 14
Top	½ x 14 x 14
Bottom	½ x 14 x 14

(Dimensions are in inches.)

will present a wider package and move across the fence with greater stability than a single narrow board.

When cutting cross grain, use a backer board to prevent tearout (see the photo on p. 94). Joint the edge of a piece of scrap to hold against the back edge of the package. Another method, which is

quicker, uses the first board in a group as the backer after it's been cut. Try this approach when you've gained some confidence on the router table.

With a package of three or four narrow boards, begin the cut and move through the first board into the second. Stop the pass with the

For a rabbet cut cross grain, two narrow boards are ganged together to pass over a straight bit. A backer board protects against tearout.

DADOES AND GROOVES

A full router pass set in from the end or edge of a board will make a cut with two shoulders, called a dado or groove. Cuts made cross grain are called dadoes; cuts made with the grain are generally called grooves (see the drawing on the facing page). These joints are also referred to as housed joints because they are often used to hold, or house, the full thickness of a board in place. This simple type of carcase work, however, can be dependent upon bit diameters matching your stock thickness. Let's first look at the ways of cutting these joints, then I'll show you how to make every joint fit even when the bit size doesn't match.

The depth of a dado or groove usually depends on the kind of job it's doing, like holding in a panel or a drawer bottom, but it should never be more than halfway through the thickness of the stock. I prefer to cut my dadoes in only one-third of the thickness, especially when they're through cuts. It's a prettier joint than going halfway with each one, and too many deep cuts can weaken a board. The strength of a dado joint depends on a tight fit between gluing surfaces that are long grain to end grain, so it's not an optimal situation. I prefer to screw my dado joints when I want added strength in a large carcase, and I cap the screws with a wood plug.

package held firmly over the bit and in tight to the fence. Then remove the first board and bring it around to the end of the group while you hold the remaining boards down. Finish up the full pass, and that first board will neatly back up the cut on the last. You'll have only one small burn mark on the shoulder of one of the rabbets to show for your pause.

When making a rabbet cut with the grain, be careful of the bit tearing out wood ahead of the cut. You'll hear a different sound than the normal one of a router bit slicing into wood when this happens—it's the sound of a router bit ripping through expensive wood and into your wallet. Take light passes to prevent this tearout, and check the grain direction of a board before routing the rabbet to locate troublesome spots. Slow your feed rate down over these spots.

DADOES AND GROOVES

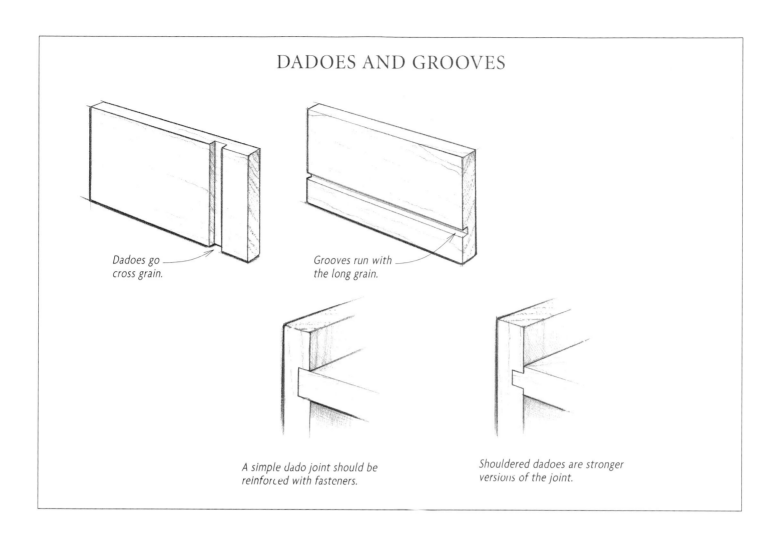

Dadoes go cross grain.

Grooves run with the long grain.

A simple dado joint should be reinforced with fasteners.

Shouldered dadoes are stronger versions of the joint.

Topside cuts with a router fence

A straight fence has limits as to how far away from an edge it can put a bit. But for cutting grooves topside, the setup and cut are very quick. Make sure that the fence is mounted with a long auxiliary fence to make it easier to keep the router lined up.

Start with the tool off the work, let the motor get up to speed, and with one hand holding the fence in tight to the work, bring the router into the workpiece. Move from left to right when working topside to help pull the router fence in tight to the work. When the cut is nearly finished, apply pressure at the end of the fence to keep the router from tipping off course.

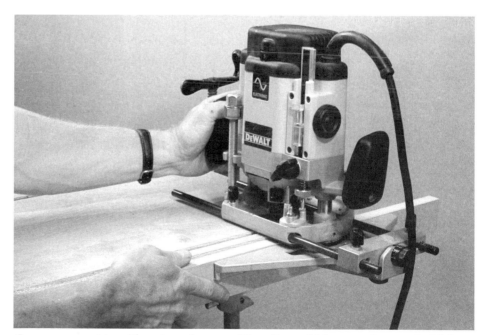

Grooves are cut topside using a long auxiliary fence. Make sure that you hold the router in tight to the edge to start and finish up a cut. At the end of the cut, also hold the fence in tight to the work.

A spacer block set against the right-angle jig extends exactly to the end of this shelf side. When it's placed flush against the end of the matching side, it will index the fence in the proper spot for the second cut.

Topside cuts with a clamped-on fence

When you need dadoes cut into a long board, a straight fence won't have enough extension to allow a cut. Using a straightedge or a long right-angle fence clamped onto the work can line up these cuts. The right-angle fence, of course, will instantly place the cut square to the edge, but you'll have to figure out the offset from the fence to the cutting edge of the bit. One useful trick is to mark out the fence with the standard offset for whatever bit and template guides you use the most.

But how do you index the matching cut? If you're working on a bookshelf, the dadoes on both sides have to be at the same height

for a shelf to line up right. One method is to lay out the joint carefully on each piece, measure and mark the bit offset required, and then clamp the fence in place. But your measuring has to be dead on.

A more accurate method is to use a spacer block made out of scrap wood. Line up your first cut and set your fence in place. Measure from the fence to the end of the board and cut some scrap exactly that size. This will index the cut for the second piece if you line it up carefully with the board's edge. Glue a fence onto the scrap piece, and you'll have a jig made for doing this dado cut in the exact same spot every time.

The risk of tearout looms every time you cut cross grain with the router. There are several strategies for coping with this. One method, and I do not kid you here, is not to worry about it. If it tears out, trim off the edge a little. But you can actually plan for this tearout by milling your stock 1/8 in. wider than the final dimension. If and when it tears out, you simply go back to the table saw and trim that extra 1/8 in. off.

A more common method is to clamp a backer board firmly next to the exit spot where you're routing. A strip of masking tape placed tight against this exit hole has also been known to work its magic in holding fibers together.

But if you're building a carcase that has a rabbeted back to it, the simplest method of all is to plan your cuts. Make all your cross-grain cuts first. If the wood or plywood tears out, keep smiling. The tearout will disappear after you come back and rabbet the back edge. In any piece with both dadoes and grooves, always cut the cross-grain dadoes first so there's never a chance of tearout in the groove.

Router-table cuts

The router table can be used to cut dadoes and grooves on small pieces. One limitation of the router table, though, is how far away you can set your fence from the bit for a dado cut. And any boards that are bowed have to be pushed flat as they pass over the bit. Narrow boards with their short ends will have a tendency to tip away from the fence. Double up these boards for better support against the fence.

Like rabbets, dadoes cut on the router table are also prone to tearout. A backer board held tight to the edge of the workpiece will prevent any problems. Always be sure you know where this exit hole is, and don't put your fingers too close to it.

Often a dado has to be cut that doesn't match any bit you have. One solution is to make two passes on the router table using a spacer and a single fence setting. A spacer block can be milled up or a piece of scrap of the right size can be used to make up the difference between a smaller bit and the final dado size needed. It fits between the workpiece and the fence for the first pass. Then it's removed for a second pass, yielding a cut of the width you want.

A dado wider than your bit can be cut on the router table in two passes. For the first pass, shown here, use a spacer block to push the workpiece away from the fence. Remove the spacer block for the second pass.

FITTING A SHELF INTO A DADO

The purpose of a shelf dado joint is to house a plank securely into the side of a carcase. The fit is perfectly tight, providing a rigid structure, but it takes some serious luck to pull this off with just one pass of a router and bit. If your bit doesn't exactly match the size of your stock, the dado won't fit and usually ends up being too loose.

My method for handling this joint is to mill my shelf stock to fit the dado cut, not the other way around. I don't try to find a bit that fits my shelves. If the stock has to be trimmed down by $\frac{1}{32}$ in. or so, that's no big deal. And if it's already perilously close to a bit size, then I'll use the next smaller bit and take two passes with it. I will either use a spacer with one fence setting (see p. 97), or I'll make a pass and move the fence over for another pass to get the dado just a bit narrower than the stock thickness.

Since most shelves need sanding or planing anyway to get rid of mill marks, you can cut the dado a hair undersize and bring the shelf stock down in thickness to fit. A hand plane or scraper does a great job of taking a board down to size. Make pencil marks across the board, and remove the entire line with passes of the plane or scraper. Try to take consistent amounts off the full width of the piece, and then check your joint. Don't sand the board to fit; sanding tends to round over ends and edges.

A shelf that fits well in a dado joint usually takes some careful thicknessing. Mill the shelf close to final thickness and then plane or scrape both faces down until the shelf fits.

Shouldered dadoes

One other way of fitting a shelf to a dado joint in a carcase is to cut both pieces to mate. A shouldered dado joint is actually the strongest dado option because the shoulders on it help to resist movement.

The dado is cut first into the carcase side. Then one or two rabbet cuts are made on the other board, holding it flat on the router table to make a tongue that fits the dado exactly. The stock must be perfectly flat with parallel faces because any discrepancy will cause a poor-fitting tongue.

If you are cutting rabbets on both sides of the board, keep in mind that each change in bit depth will give you twice that depth of cut. A way of sneaking up on a good fit is to use a paper shim underneath the board on the router table. Take a pass with the shim in place. If the tongue is still too thick, instead of raising the bit height, just remove the shim and take another pass. If you double up the paper, you'll gain another shim option. If the first pass doesn't make the tongue fit, unfold the paper for a second pass.

If your mating stock has cupped or wasn't milled with parallel sides, there's still hope for making this joint work. Use a single-shouldered dado and just make one rabbet pass holding the cupped face down to the table. This face will locate on its two edges better than the bowed side and give a better cut.

A two-shouldered dado cut can be made in cupped stock if you're careful. Use the router table and fence but hold the rabbeted piece vertically and with the cupped face

Rabbets for a shouldered dado cut can be made on the router table with the workpiece held vertically. Running only one face against the fence guarantees a parallel tongue cut. Always feed into the rotation of the bit. In this case, the board is between the bit and the fence, so it should be fed from left to right as you're standing at the table.

against the fence for the cuts. It will take two fence settings because both rabbets are indexed off of only one face, but this guarantees a tongue with parallel faces.

Make sure you pay attention to the feed direction when the board is between the bit and the fence. This cut will surprise you if you feed it in the normal right-to-left direction. The work will want to shoot out of your hands because it's cutting on the wrong side of the bit. But with calm and consideration, you can make the second rabbet cut by feeding the work from left to right or into the rotation of the bit.

STOPPED RABBETS, DADOES, AND GROOVES

Not all cuts are through cuts. Some need to be stopped short of an end or edge so that they're not visible. I did know one woodworker who through-cut everything and then patched his holes, but I think that takes just a wee bit too much effort. Stopped cuts with a router are easy and infinitely safer than stopped cuts on a table saw. The cuts will have rounded ends, but these are easily cleaned up with a chisel. And they're simple to rout before a project is glued up.

STOPPED RABBETS, DADOES, AND GROOVES

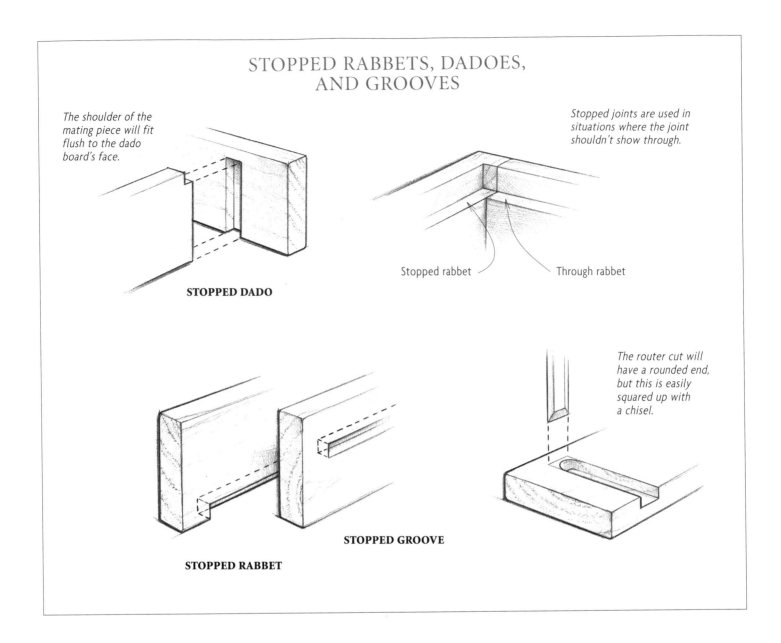

The shoulder of the mating piece will fit flush to the dado board's face.

STOPPED DADO

Stopped joints are used in situations where the joint shouldn't show through.

Stopped rabbet ╱   ╲ Through rabbet

The router cut will have a rounded end, but this is easily squared up with a chisel.

STOPPED RABBET

STOPPED GROOVE

Stopped cuts topside

Sometimes a design calls for a very clean, unobstructed look. With no dadoes showing themselves off, the eye is free to concentrate on other aspects of a design. Putting in stopped dadoes in wide boards is more easily accomplished topside. Mark out where the dado should end. I usually stop my dado cut within ¼ in. of the edge. This doesn't leave too large a shoulder to clean up on the shelf.

There are several ways to stop the router exactly short of an edge. One way is to trust in dumb luck. This strategy doesn't have a high success rate in the woodshop. But you can start your cut right over the end of the dado, plunge down into the work, and move out from there. A simple pencil mark on the wood against the router base will give you another indicator. Make this mark when the router bit is positioned right over the end of the dado. A stop can also be clamped down onto the work or even the bench if the workpiece is fixed in place (see the photo at right). The router will run up to this stop each time to be correctly located.

Stopped cuts on the router table

Most of the stopped cuts I make are done on the router table. These include stopped grooves for drawer bottoms or box lids and stopped rabbets for small cabinet backs. There are several ways of stopping these cuts, and the nature of the work will determine the right technique.

The easiest method is to not use any actual stops at all but to rely on some pencil marks and your steady hand. First mark out the cutting diameter of the bit right onto the fence. Take a square block of wood and bring it up to the bit, as shown in the photo at right. With the motor unplugged, rotate the bit by hand so that the block is pushed away from the cutter. The spot where it stops moving is the outer edge of the cutting circle of the bit. Take a pencil and mark its position onto the fence, and do this for both sides of the bit.

Mark out the cut to be made on top of your board, and line up those marks with the marks on the fence to start and finish the cut (see the photo on p. 102). Drop the board down onto the bit using those marks and pull it off when the cut is finished, but be very sure of your grip when you do this. And always start and stop exactly at your lines.

I usually prefer a more certain method of indexing my cut so I use actual stops clamped onto my fence. Without marking the bit

A stopped cut made topside can be indexed with a stop block. The workpiece is fixed in place, and the stop is clamped to the bench. Each pass of the router will end at the same spot.

Marking out the diameter of the bit on the fence will help to locate the ends of a stopped cut made on the router table. Unplug the router, hold a square block of wood up to the bit, and rotate the bit by hand to move the block over. Hold it carefully in place and mark the fence to indicate the cutting edge of the bit.

diameter on the fence, stops can quickly be set by holding up your marked-out board on edge close to the bit. Rotate the bit so its cutting edges are in line with or parallel to the fence. Eyeball the bit's edges to each end of the groove, and place stops on the fence against the far ends of the board.

With some boards and fence settings, this approach may not always be possible or yield perfect results. So another way of placing stops is to mark out the bit on the fence and lay out the stopped grooves on your board. Line up the left-side pencil marks of each, and put a stop against the far right end of the board for the first stop. Move the board down so the right-side marks align, then place a stop against the board's far-left end.

A last method is for the mathematically inclined and perfection obsessed. Mark out the stopped cut on your workpiece. Then measure out the distance from the far end of the board to the beginning of the stopped cut. Rotate the bit so its cutting edges are in line with the fence, and measure out that same distance from the far edge of the bit to the placement of the stop on the fence.

Cutting tips
Stops eliminate any chance of overcutting a groove. They also mean that only one board in a group has to be marked out so the stops can be set. Once they are, the rest of the boards will all be cut the same as the first.

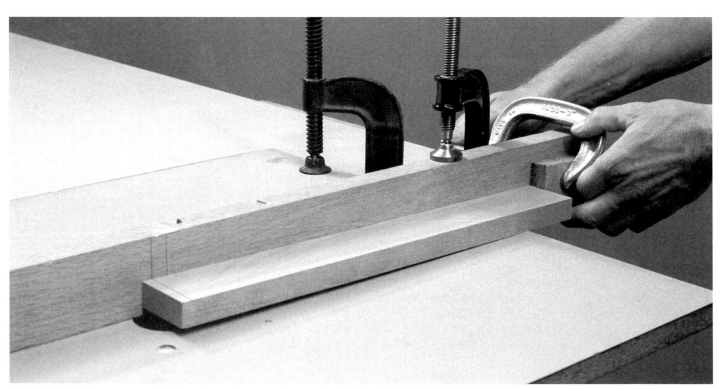

With the bit diameter marked out, line those marks up with the stopped groove laid out on top of the board. Set the stops at the far ends of the piece.

To start a cut, place one end of your board against the far right stop and an edge against the fence. As you lower the piece down onto the bit, keep good pressure against both end and edge. Make a full pass all the way to the left stop and then lift the board off the bit. Keep the board tight to the fence and end stop when removing the board as well.

Since not all router bits have center-cutting capacity, starting a cut by pushing straight down onto a bit can often produce burning. To make entry easier, slide the workpiece back and forth against the fence just a little as you're lowering it down, to get the cut started. This way the bit will cut normally and avoid burning.

Stops on a router table are likely to have chips bunch up against them as a cut is made. This debris can prevent your workpiece from properly indexing. A dust collector should eliminate the problem, but what if you don't have a good one hooked up to your router table? A simple trick will avoid this problem.

When a stop is placed, especially at the outfeed side, put a shim underneath it to raise the stop up ⅛ in. or so. I use the little wooden calipers I carry in my apron to do this. Set the stop with the shim underneath it (see the photo above right), pull the shim, and the wood dust will go under the stop or be easily blown away as you rout.

A shim placed under a stop block provides clearance beneath. Remove the shim after clamping the stop in place; the small space will allow debris to pass underneath so the accuracy of the stopped cut won't be affected.

When grooving for box or drawer bottoms, grooves for the sides and ends are usually different lengths. If they need to be cut in several passes, plan ahead. Place the stops for the long pieces first, make the passes needed and leave those stops in place. Then set the stops for the short pieces, and make the necessary cuts. Reset the bit height, cut the short pieces, pull those stops and then finish up the long pieces. If more than two passes are required to get down to depth, use one set of clamped-on stops, but only for the long pieces. Then, with spacer blocks placed against those stops, the shorter pieces can be indexed for their grooves.

DADO RABBET

This joint combines a rabbet and a dado cut. The depth of the dado must match the depth of the shoulder of the rabbet.

Dado

Be careful of the short grain created by the dado.

Rabbet

A shallow depth of cut—one-third the thickness of the stock or less—keeps the joint strong.

A narrow dado cut helps to minimize short-grain problems by keeping the dado cut farther from the end of the board.

DADO RABBETS

Another method of carcase construction is to use a dado and a rabbet joint together. The dado rabbet looks exactly like a single-shouldered dado, only it's made on the ends of two boards to form a corner rather than in the middle of a board (see the drawing at left). Like the shouldered dado, it requires two cuts, but once you've made the setup on the router table, you can cut this joint all day long for a run of box parts.

The drawback to this joint is the short grain caused by the dado cut. This can be broken off, especially in solid wood, if you're not careful. Always take care when fitting the joint, or use a very narrow dado to keep it as far as possible from the corner.

The dado is cut first, and its placement will locate the rabbeted piece. Cut it too close to the end, and the rabbeted side piece will remain proud of the corner. Set it in too far from the end, and there will be more of the end to clean up after gluing. Check your setup by taking a practice cut in some scrap and holding up a piece of the stock that will fit in the dado (see the photo on the facing page). With its outside face flush to the corner, its inside face should just line up with the dado cut.

Two depth settings on the router table

There are two ways that you can set up the router table to cut this joint. The easier way is to set both the fence and bit depth for two separate cuts with each board held flat to the table. The dado cut is made first, with the fence setting checked for accuracy as described above. A straight bit makes the shallow cuts ($\frac{1}{8}$ in. to $\frac{3}{16}$ in. deep in $\frac{1}{2}$-in. stock). This minimizes the chances of breaking out the short grain when fitting the joint.

The rabbet cut is made with the fence moved over so that just enough bit shows to cut the right-sized shoulder. Set it with too much bit exposed, and the shoulder cut will be too deep. The result is that the joint won't close up. Too little bit showing yields a sloppy-looking joint that shows a gap at its end. The bit height is adjusted to produce a rabbet that just fits into the dado. This is strictly a glue joint and so a snug fit is required. This is, of course, easy to say but takes time and patience to do.

To check the setup for a dado-rabbet cut, cut a dado into a piece of scrap, and hold up the piece to be rabbeted against it. One of its faces should line up with the dado's edge and the other with the end of the board.

Single depth setting on the router table

The other way of cutting the dado rabbet uses the same depth setting for the bit with only the fence adjusted for the rabbet cut. The dado pass is made the same, but the rabbeted board is held vertically for its cut. If you're using narrow boards that could fall or tip into a router-table hole, place a table insert down over the bit first. Some tables have removable inserts, but any piece of flat $\frac{1}{4}$-in. Masonite, hardboard, or plywood will do. Just rout or drill an access hole through it, and then set up the fence for the dado cut.

Make all the dado cuts, and then move the fence over for the rabbet cuts. Since the bit depth is unchanged, the rabbet shoulder

will automatically be cut to the proper depth. Make sure the fence is tall enough to support those boards. Set the fence so that the bit is right next to or recessed within the fence, and adjust the fence until it's cutting the right size tongue to fit the dado.

Solid-wood corner

While the dado rabbet is perfectly suitable for small boxes, larger carcases require a more substantial joint that is less prone to short-grain problems. One solution is a tongue-and-groove joint, used only in plywood construction. Here a solid-wood corner block running lengthwise through the joint is grooved to accept a rabbeted or tongued plywood side. This variation becomes a very strong application for carcase construction as well as a design element.

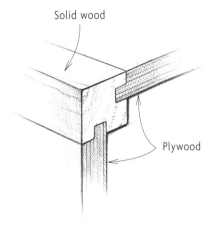

Solid wood

Plywood

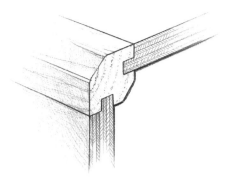

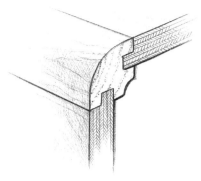

A variation on the dado-rabbet joint used for plywood construction is the tongue and groove with a solid-wood corner. This joint should not be used in solid carcase work because of variations in shrinkage.

The corner can be left square or be beveled, rounded, left proud of the plywood, or sanded flush. The inside edge of the corner can also be detailed.

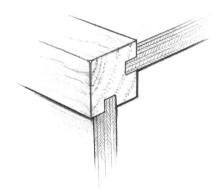

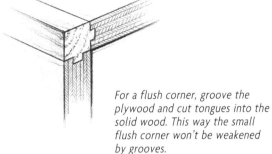

For a flush corner, groove the plywood and cut tongues into the solid wood. This way the small flush corner won't be weakened by grooves.

Large solid-wood corner blocks can be beveled or rounded off, left proud or sanded flush to the plywood. For a perfectly flush corner both inside and out, the placement of the rabbets and grooves is reversed so the grooves won't weaken a small corner block.

One word of caution—don't use this joint in solid carcase work. Because the corner will not shrink in its length as much as the solid sides will shrink across their width, this joint is not suitable for solid stock.

SMALL JEWELRY BOX

A small box will not be put to the tests that a large cabinet may undergo, so the joinery can be designed for efficiency and looks, not just strength. Dado rabbets and shouldered dadoes are perfect production-style joints. In this little box made out of cherry, I cut dadoes in the side pieces for a run of boxes. Then I set up to cut the rabbets to fit them. They were just a little tight at first, but one or two swipes with a sharp hand plane on their back sides took them down to size and cleaned up their inside faces as well.

Next I grooved all the pieces for a bottom. Putting the groove in after the dadoes are cut sidesteps that cross-grain tearout problem. Two of the grooves are stopped, and two are through cuts. A solid cherry bottom floats in the grooves. The lid is cut smaller than the box sides and then rabbeted ⅛ in. around its edges to fit down within the box. I did this with a straight bit and fence on the router table. Here again, making the cross-grain cuts before the long-grain cuts eliminated any tearout.

A small jewelry box made of cherry is joined with dado-rabbet joints, which are strong enough for this design. The ends of the box are lightly textured with a carving gouge.

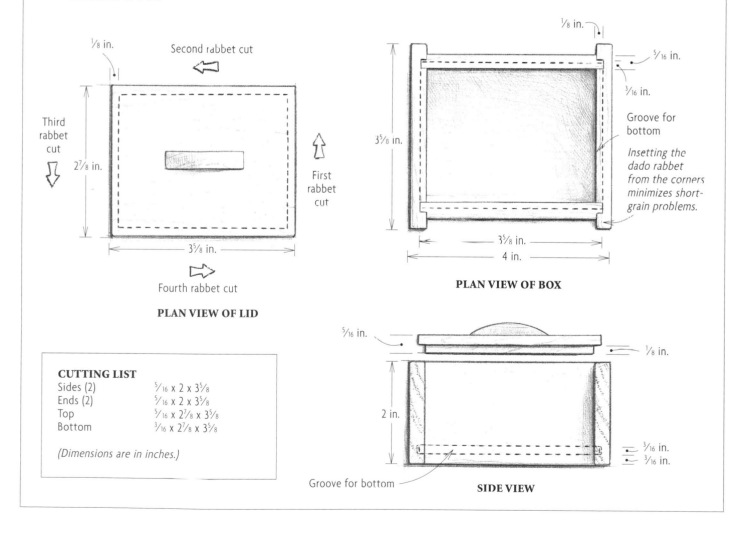

PLAN VIEW OF LID

PLAN VIEW OF BOX

Insetting the dado rabbet from the corners minimizes short-grain problems.

SIDE VIEW

CUTTING LIST

Sides (2)	⁵⁄₁₆ x 2 x 3⅝
Ends (2)	⁵⁄₁₆ x 2 x 3⅝
Top	⁵⁄₁₆ x 2⅞ x 3⅝
Bottom	³⁄₁₆ x 2⅞ x 3⅝

(Dimensions are in inches.)

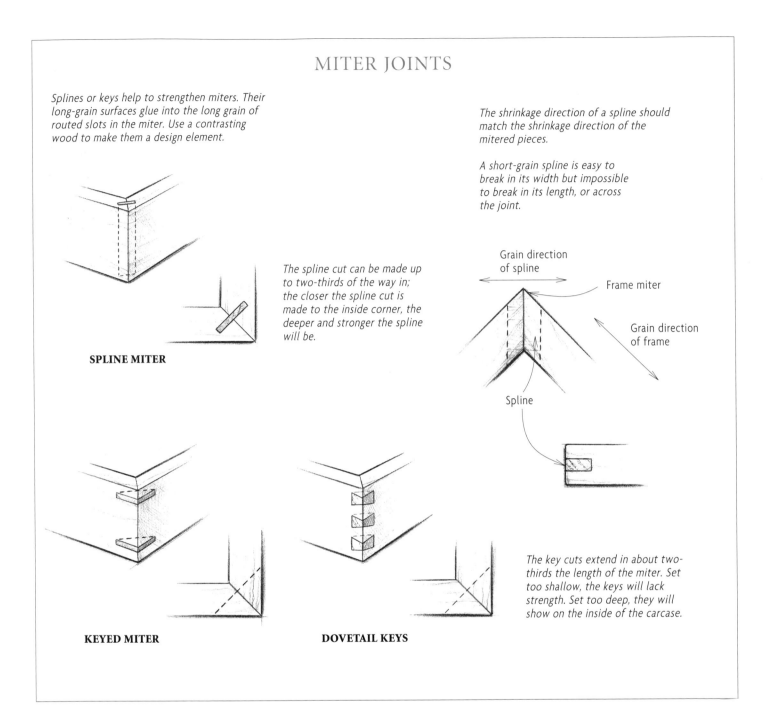

Splines or keys help to strengthen miters. Their long-grain surfaces glue into the long grain of routed slots in the miter. Use a contrasting wood to make them a design element.

The shrinkage direction of a spline should match the shrinkage direction of the mitered pieces.

A short-grain spline is easy to break in its width but impossible to break in its length, or across the joint.

The spline cut can be made up to two-thirds of the way in; the closer the spline cut is made to the inside corner, the deeper and stronger the spline will be.

Grain direction of spline

Frame miter

Grain direction of frame

Spline

SPLINE MITER

KEYED MITER

DOVETAIL KEYS

The key cuts extend in about two-thirds the length of the miter. Set too shallow, the keys will lack strength. Set too deep, they will show on the inside of the carcase.

MITER JOINTS

Miter joints are the ideal picture-frame joint because molding and rabbet cuts can be carried right off an edge. Miters are also useful for carcase work, but they're not quite a long-grain joint so they require some help across the joint. Miters can be strengthened with splines or keys (see the drawing above).

Spline miters

Splines can be added to a miter along its length. This puts long grain to long grain for added strength in gluing. The spline grooves are cut before gluing the joint. When splining in solid wood, use a solid-wood spline arranged with its shrinkage direction matching that of the mitered

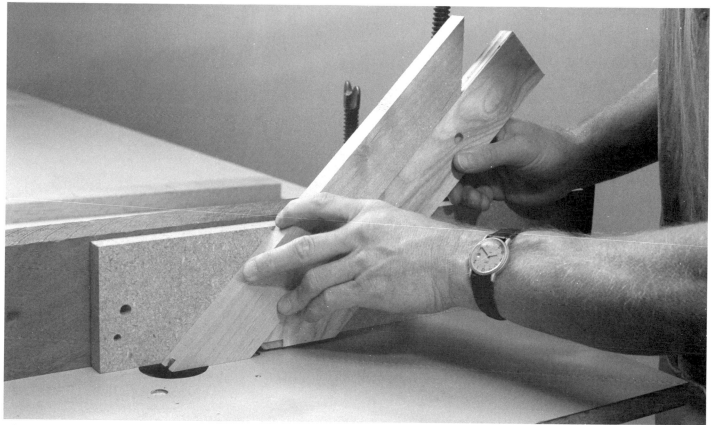

The spline-miter jig for the router table consists of a flat piece of ¾-in. particleboard 4 in. high with a fence attached to it at a 45° angle. The nails that fasten the fence are high enough up to avoid being hit by the bit. The fence also prevents tearout at the end of the cut.

pieces. This way the wood will all move in unison. It will also make the spline impossible to break across the joint. In plywood or other sheet goods, the spline should be made of plywood.

Miters can be cut with a table saw, and in box construction the length of the miter is usually pretty narrow. It's simpler, then, to cut the spline groove on the table saw, too. But in a frame with a wide miter, you can get very clean results with a router-table setup.

There are two methods for making the spline groove on the router table. The first uses a straight bit with a fence and jig. The jig is made

of particleboard or scrap plywood with a plywood fence nailed or screwed onto it at a 45° angle (see the photo above). The mitered piece sits against the jig fence for support, and the two are sent past the bit. The bit cut is centered in the thickness of the mitered pieces. If the bit isn't exactly centered for the cut, running face sides consistently toward or away from the jig always yields grooves that will line up. Just make the jig longer and add another 45° fence tipping away the bit in the other direction.

The second method uses a bearing-mounted slotting cutter recessed within a fence. Even when using

This small box in cherry has walnut keys glued into it for strength and contrast.

Keyed miters

Keys glued into miter joints provide another method of long-grain strengthening. After the glue joint has cured, cuts are made across the joint about two-thirds of the way in. Small splines or keys are fitted to these cuts and glued in. My preference is to keep keys small and neat with a slot the size of a small straight bit. However, a dovetail bit can also be used to cut the key slots, with angled keys fitted to match the dovetail angle. Not surprisingly, there are a few ways of putting all these cuts in. The first, for smaller boxes, is with a router-table cut.

The 45° splining jig shown in the photo on p. 109 can be used to support a small box for a key cut past the bit. The fence is set figuring in the thickness of the jig and does not have to be parallel to any table edge to be effective. For cuts made the same distance in from the top and bottom of the box, index first the top and then the bottom edge of the box against the jig for all the cuts. With a narrow bit, you can take several passes until you get to the right depth.

the bearing for a full depth of cut, the fence should be used for safety. If you have an adjustable fence, close up the opening around the bit and use a wide backer board to provide extra support behind the work. Miter the backer board as well. Raise the cutter so it's dead center in the mitered stock, or make sure that you run face sides all up or down for each half of the miter.

The splines are cut out of a piece of solid wood long enough to handle safely. Make sure the splines are cut accurately to length so they will allow the joint to close up during assembly. Don't worry about the ends of the splines sticking out from the joint. They can be cleaned up later after being glued in.

Make certain that when you put in the keys they fit snugly and all the way to the bottom of the slot.

Larger mitered carcases are too difficult to move safely past a bit on the router table. A template laid out with slots for key cuts is the solution (see the drawing on the facing page). It straddles the miter after the carcase has been glued together. The template is set onto the miter with angled blocks underneath to support it on the carcase. A straight bit used in a

KEYED-MITER
ROUTER TEMPLATE

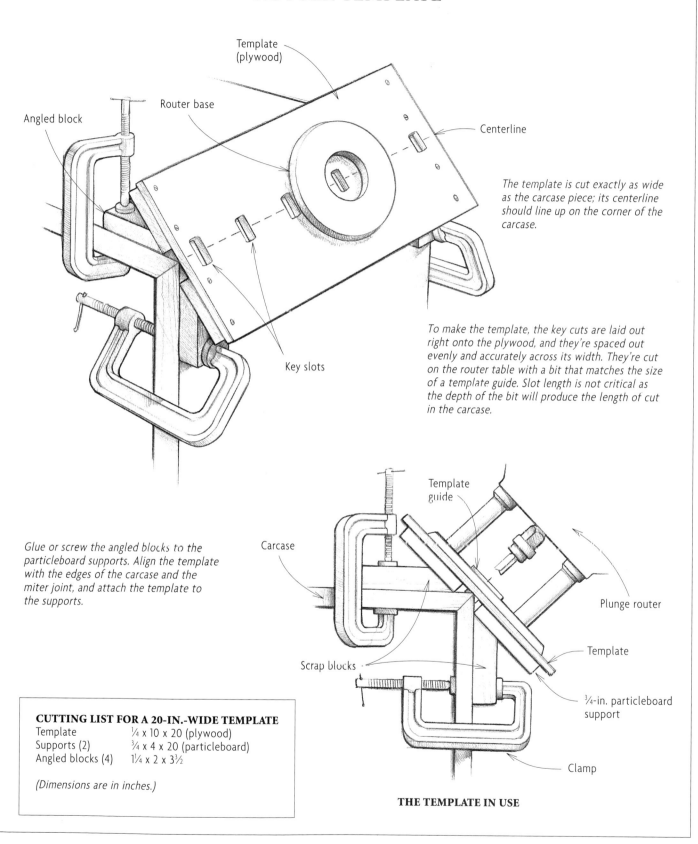

Template
(plywood)

Router base

Angled block

Centerline

Key slots

The template is cut exactly as wide
as the carcase piece; its centerline
should line up on the corner of the
carcase.

To make the template, the key cuts are laid out
right onto the plywood, and they're spaced out
evenly and accurately across its width. They're cut
on the router table with a bit that matches the size
of a template guide. Slot length is not critical as
the depth of the bit will produce the length of cut
in the carcase.

Glue or screw the angled blocks to the
particleboard supports. Align the template
with the edges of the carcase and the
miter joint, and attach the template to
the supports.

Template
guide

Carcase

Plunge router

Template

Scrap blocks

¾-in. particleboard
support

Clamp

THE TEMPLATE IN USE

CUTTING LIST FOR A 20-IN.-WIDE TEMPLATE
Template   ¼ x 10 x 20 (plywood)
Supports (2)   ¾ x 4 x 20 (particleboard)
Angled blocks (4)   1¼ x 2 x 3½

(Dimensions are in inches.)

FINGER JOINTS

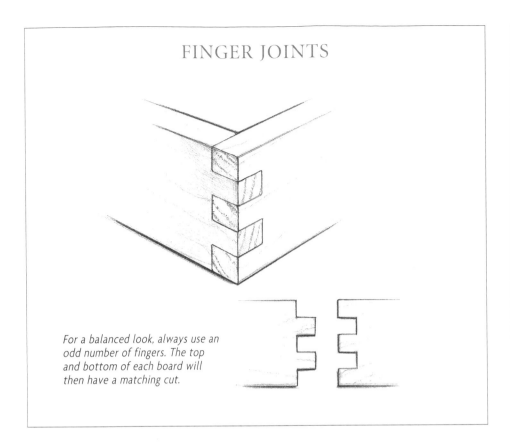

For a balanced look, always use an odd number of fingers. The top and bottom of each board will then have a matching cut.

Finger joints are often small and regular, but they don't have to be. These finger joints are part of the design of a blanket chest.

plunge router is set to depth, and the key cuts are made with a template guide riding in the slots.

FINGER JOINTS

Finger joints are designed to streamline production on a project and maximize gluing surface in a joint. Cut frequently on a table saw with a dado head, finger joints are made with the same size slot for each finger, with the workpieces offset to provide a matching fit (see the drawing above). But the router can make this joint as well as the table saw, and even better in some situations. There are two router methods I use.

Finger-joint templates

For evenly spaced fingers as wide as a router bit, dovetail jigs can be adapted to produce this joint. Only the straight bit and straight-fingered side of a jig are used to make these cuts. But this kind of box joint, like the one cut on the table saw, can be a little repetitive. You can vary the spacing in the jig, but this makes the jig less attractive as a production tool.

Router-table finger joints

For larger fingers and spacing patterns that are not so regular, I use the router table to cut the fingers. A sliding jig is set up with a fence and stop blocks to make the cuts (see the drawing on the facing page). But I first remove the waste portion of each finger slot on the

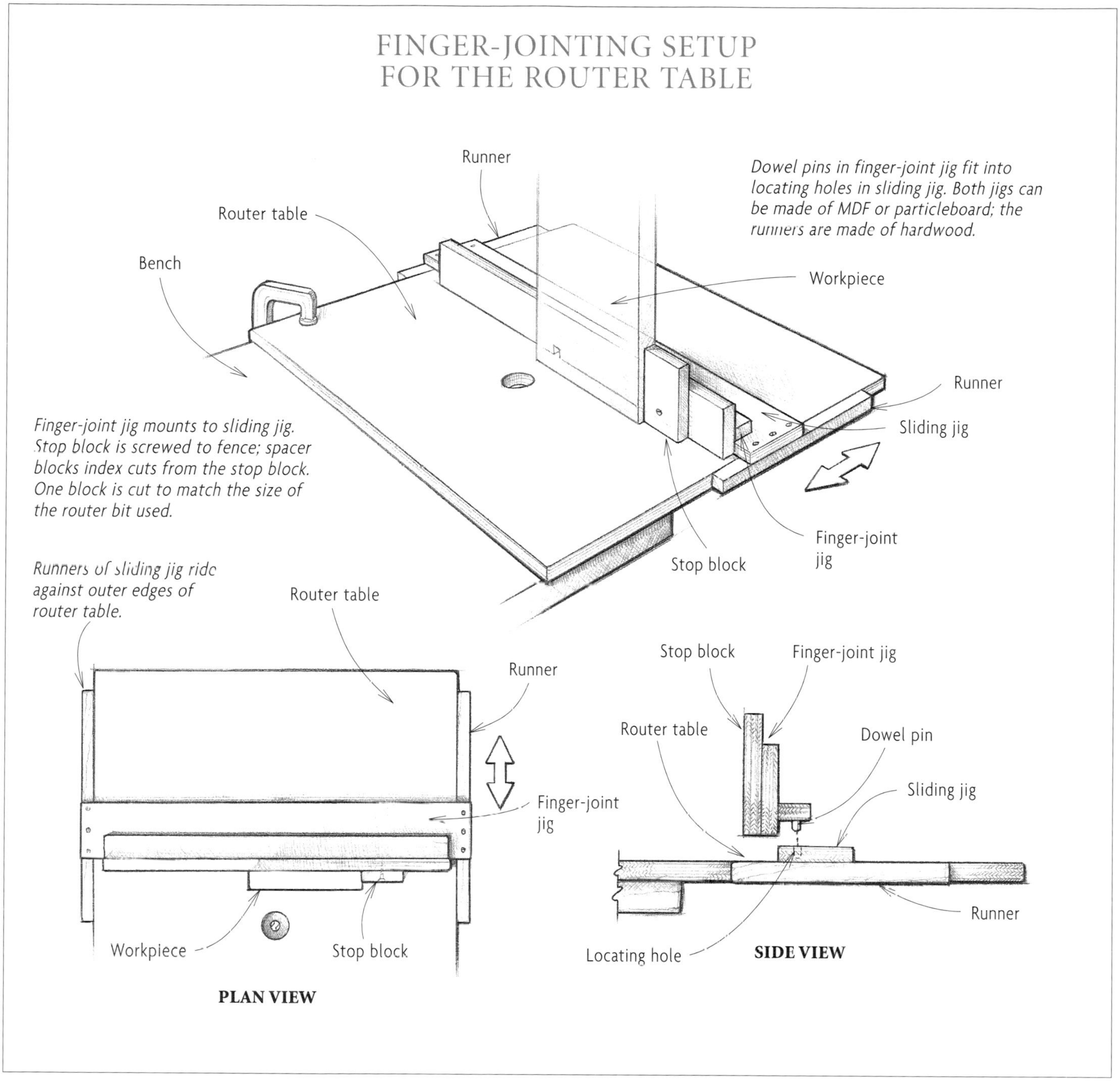

FINGER-JOINTING SETUP
FOR THE ROUTER TABLE

Runner

Router table

Bench

Dowel pins in finger-joint jig fit into locating holes in sliding jig. Both jigs can be made of MDF or particleboard; the runners are made of hardwood.

Workpiece

Runner

Sliding jig

Finger-joint jig mounts to sliding jig. Stop block is screwed to fence; spacer blocks index cuts from the stop block. One block is cut to match the size of the router bit used.

Finger-joint jig

Stop block

Runners of sliding jig ride against outer edges of router table.

Router table

Runner

Stop block

Finger-joint jig

Finger-joint jig

Router table

Dowel pin

Sliding jig

Workpiece

Stop block

Runner

PLAN VIEW

Locating hole

SIDE VIEW

bandsaw. Now my ¾-in. bit, which is my largest, still isn't as large as a finger. What the stops and spacer blocks do is establish the long-grain cut of each finger. The end-grain cut is made by carefully sliding the board back and forth over the bit.

After the first pass is made in two of the boards, their mating cut is put into the other boards. A spacer goes between the board's edge and the stop that is exactly as large as the router bit to push the second boards over. The other cuts use spacer blocks as well to move the work out from the stop. You can

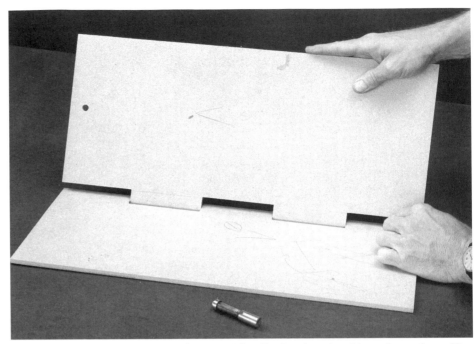

Large carcase finger joints can be cut with a flush-trimming bit and templates like these. The templates are made up to fit each other perfectly. The bit cuts exactly to the template so your workpieces will end up fitting just as well. The corners of each finger will have to be cut or chiseled square after routing.

play with the fit of the fingers by shimming between the spacer blocks with paper or tape. Always make practice cuts in scrap before committing your good wood to a setup.

Templates for large finger-jointed carcases can also be cut on the router table using this method. When the templates fit each other perfectly (see the photo above), they can be used to index these finger joints using a flush-trimming bit in a plunge router held topside. Attach them with double-stick tape to the workpieces. The inside corners of each finger end up round from the router cut and get cleaned up with a saw or chisel.

SLIDING DOVETAILS

There's been more hair lost to the frustrations of fitting a sliding dovetail than probably any other joint cut. At least those with hair have something to yank on. But sliding dovetails are such a great joint for drawer and carcase construction that they're worth the effort. They're mechanically strong, hold boards flat, and tie carcases together. Here are some ways of making this joint work.

Female cut

For drawers, boxes, and other small pieces all the cuts will be made on the router table. If you need a table insert to cover a large bit hole, put it in before starting. The female part of the joint is cut first, and it has to be cut with the dovetail bit at full depth because of its angles (see the drawing on the facing page). But save your

ROUTING A
SLIDING DOVETAIL

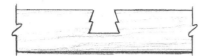

You can't take a series of passes with a dovetail bit because it would make for a stepped cut. Instead, use the two-step approach shown below.

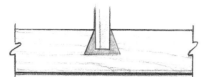

Step 1: Use a straight bit in a series of passes to clean out the waste.

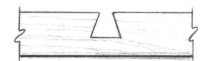

Step 2: Use a dovetail bit at full depth on the final pass.

dovetail bit and your ears by cutting out the waste with a straight bit. A ½-in. dovetail bit is wide enough at its narrowest to allow a ¼-in. waste pass first.

Set up the fence distance on the router table with the dovetail bit mounted. Then remove the bit and install a ¼-in. straight bit almost to the full depth of cut. The fence setting remains because the straight bit is centered in the same place as the dovetail bit. Make all your straight cuts in the pieces receiving the female part of the joint. Make the dovetail pass with the bit set to full depth (see the photo at right), and check that the cut has a consistent depth throughout. If this is a stopped cut, use a stop for both the waste pass and the dovetail cut. At the end of a stopped cut, let the motor slow all the way down before pulling the work back off of the bit.

The groove for a sliding-dovetail joint can be cut in two passes. After a pass with a straight bit to remove much of the waste, the full cut can be made with the dovetail bit.

To sneak up on the male cut of a sliding dovetail, put a paper shim between the workpiece and the fence. This kicks the piece just a few thousandths of an inch away from the bit, so the tail will be cut that much bigger. If the joint still doesn't fit, remove the shim and take another pass.

Male cut

The tail piece is made with the piece held vertically against the fence. Keep a firm grip or use a tall fence for good support because the work cannot be tipped into the bit. Hide the bit in the fence recess so only a little of it is showing but do not change its height setting! This keeps life simple and eliminates one of the frustrations of this joint. Then take a pass on a piece of scrap milled to the exact thickness as your stock, flip it face for face and cut the other side of the joint. For thinner rails being cut with a dovetail, support the rail with a backer board as it passes the bit.

Keep cutting and checking the fit as you go. The fit we're looking for is one that allows the joint to go just halfway together with hand pressure. It will have to be carefully

tapped apart. Check both sides of the tail piece too by flipping it around and sliding it into the joint. A fit that allows the two pieces to go all the way together is usually too loose, and glue won't help.

If the tail is still too large, not enough bit is showing. Pencil-mark the fence position down on the table, loosen just one of the clamps and tap the fence over. With more bit showing, more material will be removed. And with each new setting, there are two cuts to be made. Keep checking the fit and breathing deeply.

If the joint fits together sweetly the first time, give thanks and go play the lottery. But one final method to sneak up on this fit is to use shims against the fence (see the photo at left). Say your fit is almost there, but you've moved the fence over for one more pass and think it will cut too deep. Grab a piece of paper or a dollar bill (or a hundred-dollar bill if you want a really thick shim) and put that between the fence and the workpiece. This will kick the board over by just a few thousandths of an inch and give you a bigger tail. If it's still too large, remove the shim and take the final pass.

Single-shoulder sliding dovetail

There's no stronger choice for a joint than a dovetail (see the drawing on the facing page), but fitting them can require gobs of patience at your router. But if you make one side of the joint flat, you can do your fitting with a hand plane rather than with the router.

The single-shoulder dovetail is cut the same as the regular sliding dovetail except that the straight

DOVETAIL JOINTS

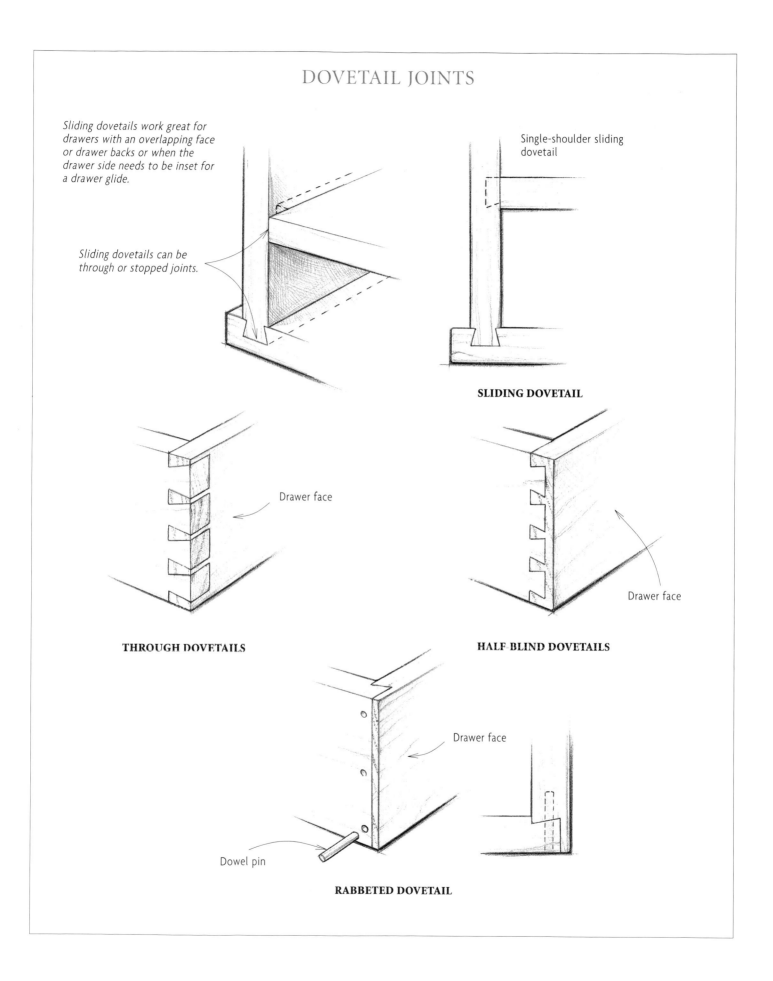

Sliding dovetails work great for drawers with an overlapping face or drawer backs or when the drawer side needs to be inset for a drawer glide.

Single-shoulder sliding dovetail

Sliding dovetails can be through or stopped joints.

SLIDING DOVETAIL

Drawer face

THROUGH DOVETAILS

Drawer face

HALF-BLIND DOVETAILS

Drawer face

Dowel pin

RABBETED DOVETAIL

DOVETAILED TOOLBOX

I carry around my carpentry tools in a nailed-up plywood toolbox. It suits the purpose. I made the handled toolbox shown in the photo at right to display some other very special tools. The dovetails are strong enough for the job, and I could really show them off with the contrasting oak and poplar that I used. The box is made with ¾-in. stock and measures 5¾ in. by 16 in. by 18 in. I milled the stock to size and marked the face side of each board.

A handled toolbox of oak and poplar shows off the beauty and strength of dovetail joinery.

The tails were cut first. I put the dovetail bit in my fixed-base router and set its depth for just under the combined distance of the template and the thickness of the stock, 1¼ in. A gauge line set for slightly less than the thickness of the stock marked out on the boards gave me another guide for checking this depth. Then I put a board in my vise with its face side away from me. I used a spring clamp to attach a piece of scrap to the outside face so my clamps wouldn't mar this surface.

I centered the tail template on the board by measuring it so the ends of the board were set in from a tail finger an equal amount. I put the clamps on low enough to stay out of the way of the bit. The cut was made holding the router flat to the template with no tipping or rocking. The bearing on the bit fits perfectly between the fingers.

Pins are marked out from the tail board using a scratch awl. Here, a hand plane under the tail board raises it off the bench top.

When all the tails had been routed, I marked out the pins. This was done with the pin board low enough in the vise so that the tail piece could be laid on top of it. Then I reversed the marked pin board in the vise so its face side was out or away from me again. This was a critical placement so those marks would line up with the fingers of the pin template. Also by placing the pin board this way, the waste shot out away from me when I was cutting.

The flush-trimming bit that comes with the jig cut the pins. I used a light scoring cut first on the outer edge of the board. Then I moved the bit into the finger slot. The depth was set the same as for the tail cut. Before removing the first pin board, I placed a stop against one edge to index all the other pin boards.

The toolbox is grooved to hold a ¼-in. plywood bottom. Two of the cuts are stopped, and two are through cuts. These were all made on the router table. The slots for the handle were cut after the box was together. Finally, the handle was glued and screwed in place. A fair amount of rounding on the handle's bottom edge give it a welcome feel. Three coats of oil, and I had a nice little toolbox.

One other issue pops up, and that is the width of your stock. Some boards will end up with half-tails using the templates. I prefer half-pins at the outer corners of my joint because this is a stronger option. The point at which you need the greatest strength is down low, which in a half-tail is at its narrowest. So I avoid half-tails by using either larger half-pins at the outside or cutting the stock down in width.

HALF-BLIND DOVETAILS

When the strength of a dovetail joint is required but you don't want to see the joint from one side, the half-blind dovetail (also called the lapped dovetail) can be used. Typically employed in drawer construction, this joint has its tails and pins arranged so the angle of the tails opposes the movement of the drawer. The side pieces are cut with tails; the drawer front, with pins.

There are several commercially available jigs that cut these dovetails. They produce evenly spaced identical dovetails with varying results. Their success depends upon careful setup of the bit and jig. Both boards of this joint are cut at the same time.

Place the tail board in the jig first but fairly high inside the clamps. The purpose of this is to provide a stop for the pin board to butt up against. With the pin board located and clamped into place, then move the tail board down to line up its end grain exactly with the long grain of the pin board. When the alignment is perfect, place the finger template flat on top of the pin board. A template guide will ride within these fingers.

Mount the guide in your router base and then fit the proper dovetail bit through it and tighten it in the collet. The depth of cut controls the fit of these tails so it's critical to get this right. If it's set too deep, it cuts too loose. Set for too shallow a cut and the joint won't fit together at all. There's really only one setting at which the fit is just right, and several practice cuts will have to be taken before this fit is achieved. Cut a block of wood with a dado in it to fit exactly over your dovetail bit to index your bit height for the next time you cut this joint.

A first scoring pass is made against the entire face of the tail board. Then the bit is brought in to cut to the full depth of each finger making very sure that the template guide moves smoothly around and into each finger. I do a final clean-up pass to ensure that no section has been missed. As with the through-dovetail jig, try to end up with half-pins at the outside corners for better strength.

Half-blind dovetail joints show from one side only. When used in a drawer, the drawer side shows the dovetails. In this way, the dovetail joints will resist the pull of the drawer each time it's opened.

For router-cut half-blind dovetails, the pin board goes face side down in the jig; the tail board is clamped face side in. Locating pins on the jig automatically offset the boards so they can be cut simultaneously with a template-guided dovetail bit.

A frame with half-lap joints is clamped together. Bar clamps pull the frame members into one another, and C-clamps pull the half-lap joints tight.

router base, but you'll have to make sure the weight of the fence doesn't pull the router base down into the work. It's not really that heavy, but when your attention wanders, gravity is always ready to grab it. Use a longer auxiliary fence on a straight fence to provide better support against the ends of the frame pieces.

The right-angle jig (see pp. 43-44) can also be used to make the shoulder cuts square to an edge. If the router-base edge rides against the fence, then the distance from base edge to cutting-bit edge needs to be measured out and marked onto the work.

When using the router to cut half-laps topside, the length of the joint can present a problem. With wide frame members and a wide joint, the router base becomes more difficult to support out at the end of a board. Offset bases (see p. 42) can be used if you're careful by using sufficient pressure on them to keep the router from tipping into the work. Another method of supporting the router base is to set up a piece of scrap near the end of the frame pieces being cut (see the top photo on the facing page).

clamp pads (see the photo above). The half-lap surfaces and edges can then be cleaned up with a hand plane or sander.

In a frame with members the same size, the joints are all cut identically so the setup time is quick. Set the bit depth for just less than half the thickness of the stock. I bandsaw off most of the waste first and use the router to make the final cut. Ganging several frame members together will increase the support for the router base and provide backup behind all but the last piece in the group. Put a backer board behind that, and the cross-grain shoulder cut can be made without fear of tearout. Use the widest bit you have to make these half-lap cuts. The cuts will take less time and come out smoother.

As usual with the router, you need to select a method of directing the cut across the board or boards. A fence could be mounted to the

Router-table cuts

When half-laps get too long (over 2 in.) to cut safely topside, the router table is easily set up for the job. You can set the fence distance using one of the frame pieces as a gauge. Bring one piece up to the fence with the bit rotated so that one cutting flute of a straight bit is at a point farthest from the fence. Have one end of the fence loosely clamped so that it can be pivoted in place. Set the fence for just under the width of the board.

As with topside half-laps, it's simpler to remove most of the waste on the bandsaw before starting to rout. One rip cut against a fence and one crosscut close to the final distance of the shoulder will do the trick. Save the offcuts to use later as clamping pads.

If you don't have a bandsaw that can do this job, the half-laps can still be cut, but at a more meditative rate. Use your largest-diameter straight bit and set the bit height at about 1/8 in. Then start to nibble away the material. I like to double up two boards at a time for better support (see the photo at bottom right), and I keep resetting my bit height until I get to full depth.

Tearout can occur on the back edge of the rear board, on both the long grain and the end grain. A trim pass against the back edge of one frame member removes any long-grain tearout; a backer board will protect the crossgrain cut. Never have your hands cross over the bit or place them at the end of a cut. Feed right to left, and take your time.

As you finally get to the shoulder cut you'll be taking a bigger bite out of the wood. You'll be able to hear this in the sound of the motor. Keep your feed rate steady past the bit when making this end-grain cut. Too fast a cut, and you'll get a ragged or inaccurate cut. Too slow a rate, and the bit will burn the end grain.

After making all the cross-grain cuts, clean up the half-laps to get any waste that might remain. Move the boards in and out to the fence, with the fence acting as a stop. There will probably still be a few little ridges left to clean up with a rabbet plane or a wide chisel.

The right-angle jig directs the cut when routing half-lap joints topside. Support the far side of the router base with a piece of scrap milled as thick as the frame pieces. That will eliminate the danger of the bit tipping into the work.

Half-laps on the router table should be cut with a series of passes, using a backer board to prevent tearout.

HALF-LAP MIRROR FRAME

Half-lap joints in a frame provide a lot of gluing surface and strength. After they're cut, you can shape a curve or angle into them and still leave plenty of gluing surface for the joint.

I first drew this mirror frame out full scale on a piece of brown butcher paper (see the drawing below). This gave me the chance to look at the shape and proportion of the piece before committing my wood to a saw. The curved members I drew in by laying on and bending a thin piece of wood against the drawing (see the photo at top left on the facing page).

When the curve looked good to me, I made a template of it. Templates serve a couple of useful functions. First, they give you a pattern to use again. But more important, it's easier to cut and shape a curve onto template material (I use ¼-in. hardboard or MDF) than to shape it into ¾-in. stock. For this mirror, I redrew my curves on the hardboard, rough-cut them on the bandsaw, sanded close with my homemade spindle sander, and then cleaned up the curve with a spokeshave.

Because of the large size of these half-laps, I cut them on the router table so they had better support. But before I routed, I trimmed off the waste on the bandsaw. I then set my widest bit at slightly less than half the thickness of the stock. The fence was set to cut only 2½ in. deep into each frame member. Only if you cut your frame members as long as the full outside dimensions of the mirror frame do you cut a full 4-in.-square half-lap.

I shaped the inside curves before gluing but left the outside edges flat for clamping purposes. The inside shape was marked out and rough-cut on the bandsaw. Then I attached the template to the face of each board with double-stick tape and made the finished inside curved cuts on the router table, using a flush-trimming bit with a bottom-mounted bearing. The

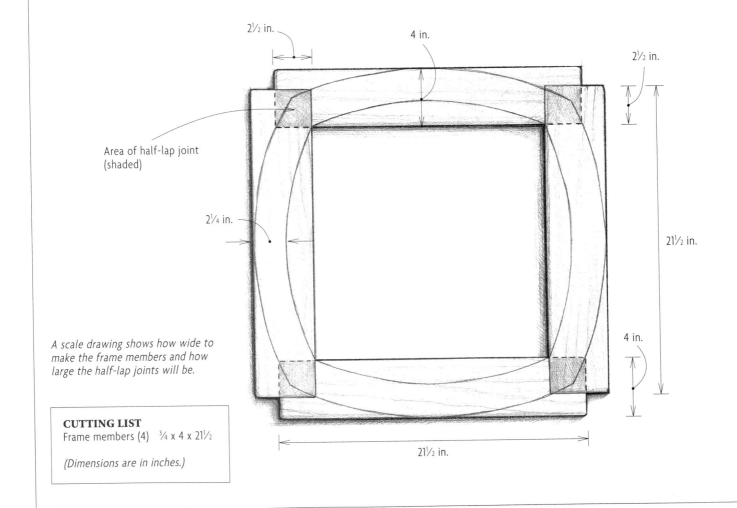

A scale drawing shows how wide to make the frame members and how large the half-lap joints will be.

Area of half-lap joint (shaded)

2½ in.   4 in.   2½ in.

2¼ in.

21½ in.

4 in.

21½ in.

CUTTING LIST
Frame members (4)   ¾ x 4 x 21½

(Dimensions are in inches.)

To draw a curve onto a full-scale drawing, bend a thin strip of wood to use as a guide.

A square frame should have diagonals that match each other in length.

bearing rode against the template while the bit cut the curve exactly to shape. I stopped just short of the ends to prevent tearout. A file and spokeshave cleaned up the ends, and a scraper and sandpaper finished the job.

The frame was dry-assembled before gluing to check the fit of the joints and to get the clamps ready. Long bar clamps pulled the shoulders in tight. One set went on top; the other set went beneath the frame, raising it off the bench so the C-clamps could be applied. Offcuts were used as clamp pads to protect the wood and to spread out the pressure of the C-clamps. I also checked for squareness by measuring both diagonals (see the photo above right). If they aren't the same length, you need to pull in the the longer direction with a clamp.

After gluing, I flattened and sanded the frame surfaces. Then I used the template again to mark out the outer curves and rough-cut them on the bandsaw to get close to the line. A template and a flush-trimming bit finished up the job. You can use the bottom-mounted-bearing bit again or use a top-mounted-bearing bit to cut the shape. If your frame is too large to run on the router table, a flush-trimming bit with a top-mounted bearing in a hand-held router (see the photo at right) can do the job in a few passes.

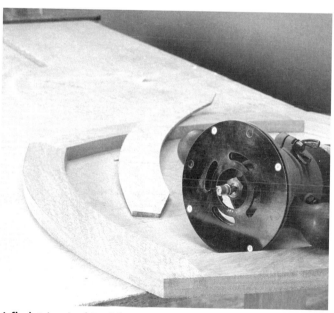

A flush-trimming bit with a top-mounted bearing routs the outer curve in two passes. On the first pass, the bearing rides against the template. On the second pass, the bit is lowered so the bearing rides against the material already trimmed.

MITERED HALF-LAP JOINT

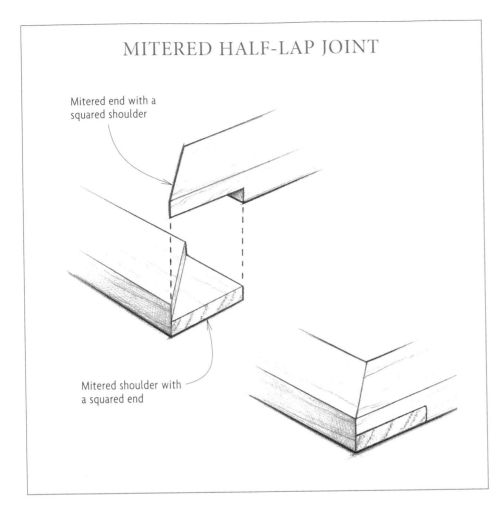

Mitered end with a squared shoulder

Mitered shoulder with a squared end

Mitered half-laps

The beauty of a mitered frame is that you can shape its inside edges straight through with a decorative cut. But a miter of any decent size requires some added strengthening. The mitered half-lap frame is a good compromise. It offers the shaping qualities of a miter along with the strength of a half-lap.

The frame members of a mitered half-lap are not cut the same. One end gets a squared shoulder with a mitered end; the other, a mitered shoulder with a squared end (see the drawing at left). Square or mitered crosscuts are most easily made with a backsaw or on the table saw. The half-laps can be cut with the plunge router and a straight bit. Make up a 45° fence and clamp this onto one frame member to cut the mitered shoulder (see the photo at left). The cut is just shy of halfway through the thickness. The board with the mitered end uses the right-angle jig laid onto it and offset the proper amount to cut the squared shoulder.

Cutting a mitered half-lap. A piece of plywood cut at 45° acts as a fence to run the template guide against. A block glued or screwed onto the plywood parallel to its long edges lines it up each time for a cut.

Cross half-lap joint and T-halving

Half-laps don't always land on the ends of boards. Sometimes they provide structural support in the middle of a span (see the drawings on the facing page). A cross half-lap is cut into both boards to form a cross, using the same right angle setups. The T-halving needs a slot in one board and a half-lap in the other to fit. For either joint, set the router to cut almost halfway through each board. The right-angle jig indexes the cuts for the slots across each board.

T-HALVING AND
CROSS HALF-LAP JOINT

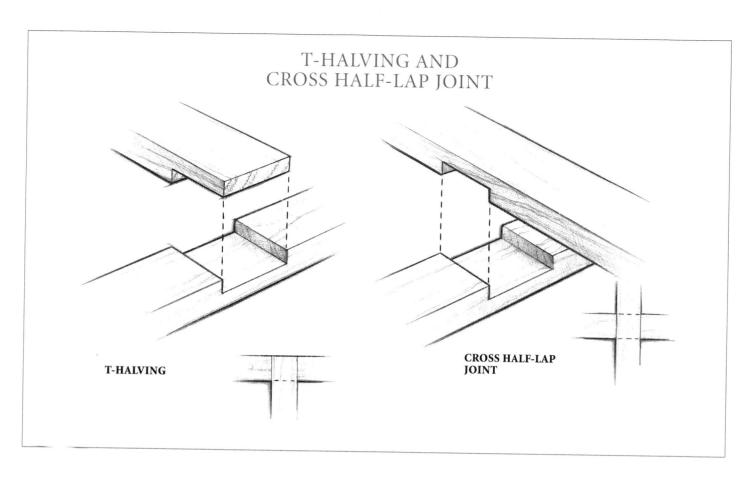

T-HALVING

CROSS HALF-LAP JOINT

DOVETAILED
HALF-LAP JOINTS

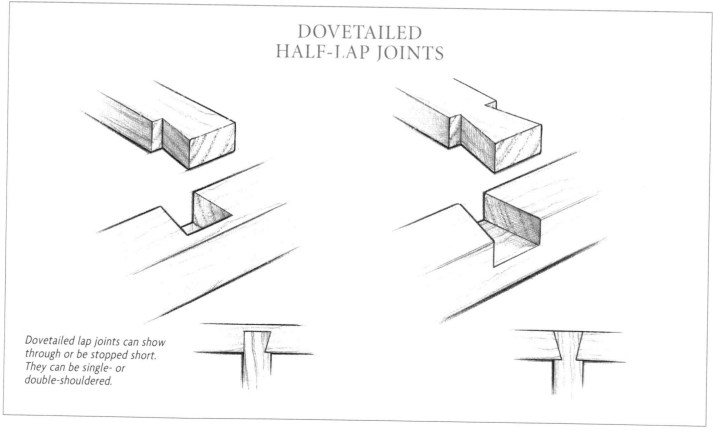

Dovetailed lap joints can show through or be stopped short. They can be single- or double-shouldered.

The shoulders of a cross half-lap joint should fit tight in order to look good and function well.

With half-lap joints like these, you can cut and fit, cut and fit, and then all of a sudden it's too loose. So don't try to set up the cut to be exactly the width of a piece—I lay out the sides to be slightly less than the width of each member. This joint relies on the long grain for most of its strength, but the shoulders of it help to resist movement as well. Its looks depend upon how tight the fit is between the end grain of the shoulders and the long edges of each board (see the photo above). Cut the slot very close to fitting and then, since you have to clean the mill marks off the edges anyway, take a hand plane and trim the board down to fit.

Dovetailed half-lap joint

When a project calls for additional strength in the middle of a board, consider using a dovetailed half-lap. This angled shouldered joint can be cut in two ways. I find it simpler, however, to cut the dovetail first, and then mark out and cut the slot it fits into.

The angles of the male part of the joint can be cut by hand or on the bandsaw or table saw. The half-lap is cut topside or on the router table. Just as with hand-cut work, the dovetail is marked onto the other board. Butt the shoulder up against the board, and mark the joint using a marking knife or scratch awl. Even if both angles of the dovetail aren't exactly alike, it doesn't matter, since you are transferring their outline to the mating piece.

Although a router jig can be designed to cut the slot for this dovetailed half-lap, I prefer to cut the shoulders of this joint with a saw and then just rout out the middle (see the photo on the facing page). The router doesn't have to make every cut in the shop. Just make a freehand cut close to the shoulder lines. The router offers a great way of getting the depth consistent. The wood that is left uncut is easily cleaned up flush to the rest of the joint with a wide chisel.

Freehand routing cuts the dovetail slot to a consistent depth; a chisel removes wood that remains close to the shoulder.

BRIDLE JOINTS

Bridle joints are similar to mortise-and-tenon joints (see Chapter 9). The difference is that the mortise is always a through cut made on the end of one board with a tenon cut to slip into it. Also called slip joints or slot mortises, bridle joints provide excellent gluing surfaces and strength. A mortise cut that's less than one-third the thickness of the stock looks best, but let it match a bit size for simplicity. Used in the middle of boards, bridle joints provide structural support for longer members. Various bridle joints are shown in the drawing on p. 132.

Slot mortises on the router table

A slotting cutter would seem to be the bit of choice for cutting this through mortise on the end of a board. But these bits have a small depth of cut so that their large diameter is relatively safe to operate. Since most slot mortises require cuts the full width of frame members, cuts that are 1 in. or more in depth are common. The table saw is the best tool for cutting this joint. But if all you have are a bandsaw and a router, you can still make this joint if you take some care.

The safest way of getting this much depth of cut on the router table is by bandsawing out most of the waste first. Set up the bandsaw fence to cut the cheeks, or pencil-mark the joint and rough it out. Get as much of the waste removed as you can so the bit will make cleanup passes in the mortise. Use a bit with a ½-in. shank, as there is a lot of stress on the bit in this cut.

BRIDLE JOINTS

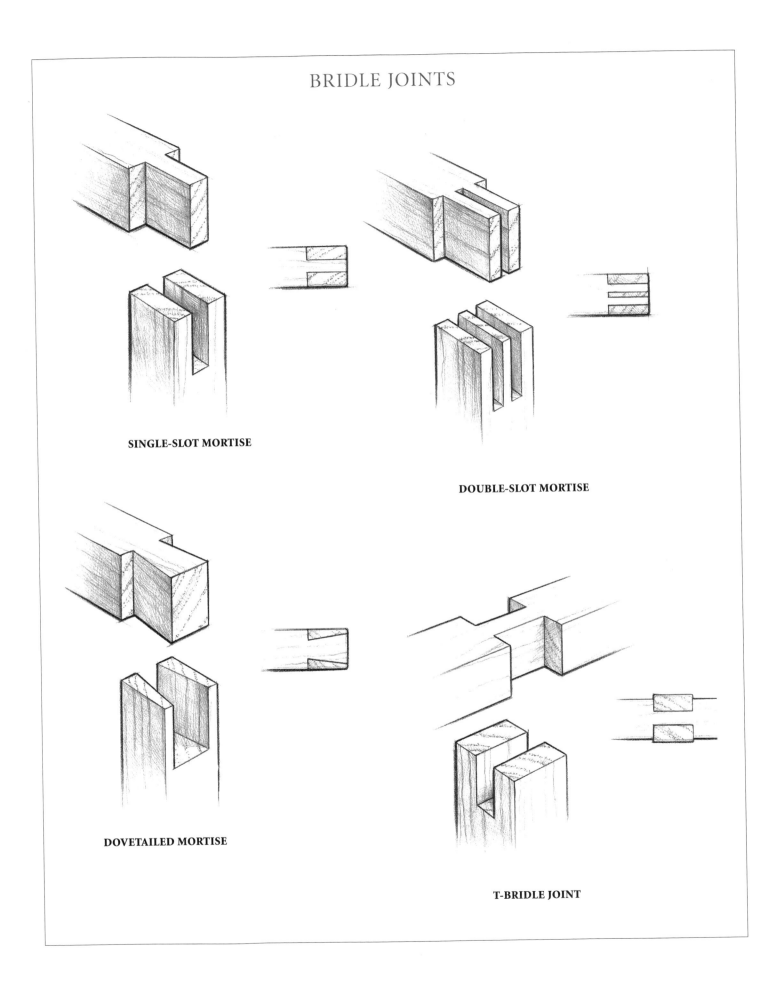

SINGLE-SLOT MORTISE

DOUBLE-SLOT MORTISE

DOVETAILED MORTISE

T-BRIDLE JOINT

A simple jig is used to support the work on the router table as it passes vertically by the bit. It's made of ½-in. plywood with a support block attached to it at a right angle with screws (see the photo at right). Make sure the screws are placed high enough that the bit won't run into them.

To provide smooth running across the bit hole, make sure there's a table insert over it. Then set the fence distance, figuring in the thickness of the jig. Take several passes to get to the full depth of the slot. If the support block is low enough to back up the cut, it will eliminate any chance of tearout. If it is not, just add a backer board behind the piece being cut.

The tenon for the joint is cut like the half-lap, flat on the router table. But there are two cheeks and shoulders to cut instead of just one. Trimming the waste on the bandsaw makes life simple. When setting the bit height for this cut, remember that your stock must be milled flat and parallel in order to come out with an accurate tenon. Try to achieve a fit that slips in with just a little bit of hand pressure. Too tight a fit will make the sides of the mortise bulge out.

Clamp this joint together just like the half-lap joint, pulling the tenon shoulders down and into the mortise. Also clamp across the joint to pull the cheeks in tight.

T-bridle-joint cuts

Bridle joints can be a headache because there are three things to figure out and fit all at once. To maintain some semblance of sanity, just work on one part of the joint

After most of the waste has been removed with a bandsaw, the mortise is cleaned up in a series of passes on the router table. A jig supports the workpiece vertically and backs up the cut. Wrap your hand around the workpiece and hook onto the jig to hold the work tight. A table insert covers the bit hole so the workpiece won't fall in.

at a time, starting with the slot at the end of the board. Cut it just like the slot mortise (see the previous section) or use a table saw. Clean it up and get it perfect.

Next, cut the large dadoes of the bridle-joint tenon. Mark out their position on the board and the thickness of the tenon that will fit the slot mortise. Set the bit height on the router table to yield that thickness. Remember that you'll be cutting on both sides of the workpiece. Test your settings by cutting into a piece of scrap the same size as the workpiece to get this fit just right.

To cut the shoulders of a T-bridle joint, use an auxiliary fence that is long enough to clamp a stop onto and index the shoulder cuts with a spacer.

A half-lap joint cut into the width of the boards is liable to break if struck.

Now concentrate on the fit of the shoulders of these dadoes. They need to be lined up perfectly to have a tight-fitting joint. To cut them, use a crosscut jig on the router table with a long auxiliary fence and a stop (see the photo at left). But unless you have a bit as wide as your stock you'll need to index and cut the second shoulder too. You can use a spacer block against the first stop to move the board down, but the spacer has to be accurately cut.

Another way is to set the first stop, remove the waste in the middle of the joint, and then move the stop over to make the second shoulder cut. Do this just under the width of the slot-mortised board, and then use a hand plane to bring that board down to the perfect size and fit.

Strengthened half-lap joint

When a half-lap joint is cut into the width of a board rather than its thickness, a potential problem arises. With so much material removed, the long grain becomes weak and susceptible to breaking if struck right at the joint (see the photo at left). Fortunately, there is a way to strengthen this joint. The joint is called the strengthened half-lap joint (see the drawing on the facing page), and it has shoulders at the weakest points of the half-lap joint that take away any flex or give there. This prevents the joint from snapping or breaking away at that spot.

The strengthened half-lap joint is cut like a bridle joint with the crosscut jig on the router table. Simplify the process by cutting one side of the joint first and measuring and marking out onto the next piece for the final cuts.

STRENGTHENED HALF-LAP JOINT

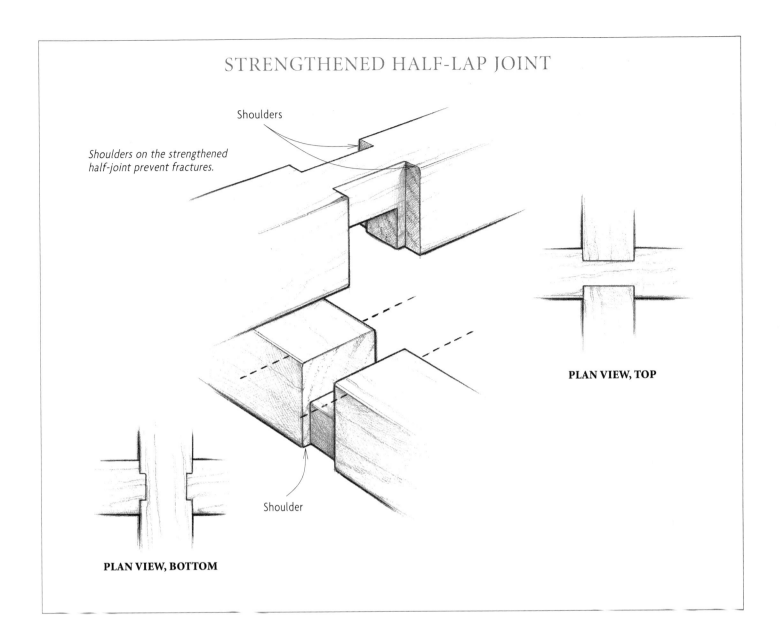

Shoulders on the strengthened half-joint prevent fractures.

Shoulders

Shoulder

PLAN VIEW, TOP

PLAN VIEW, BOTTOM

TONGUE-AND-GROOVE JOINTS

Tongue-and-groove joints (see the drawing on p. 136) provide extra gluing surface and strength across a long-grain glue joint. With the superior adhesives we have today, strengthening these joints isn't often required, but in some situations these joints can help. They are also used to form corners in plywood carcases or to apply a heavy edge banding.

Router table with a straight bit

Grooves are easily cut on a router table with a fence setup. A straight bit makes the cut on the edge of a board in either one full pass or in a series of smaller passes for a deeper groove.

Make sure that all boards have a flat edge or that longer ones get pushed flat as they pass by the bit. To ensure a centered groove, make one pass and then flip the board

TONGUE-AND-GROOVE JOINTS

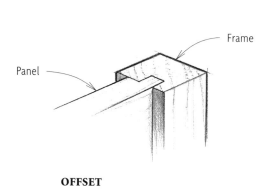

Groove

Tongue

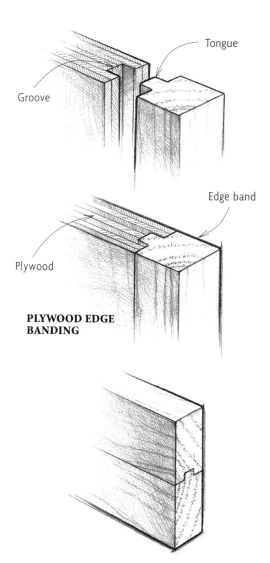

Plywood

Edge band

PLYWOOD EDGE BANDING

LONG GRAIN TO LONG GRAIN

Panel

Frame

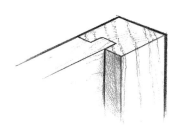

OFFSET

FLUSH

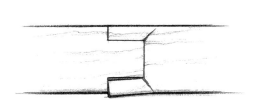

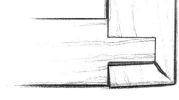

Tongues set too deep or too close to a face may break or crack the grooved piece.

face for face, and make another pass feeding into the rotation of the bit again. This gives a cut that is an equal distance from both faces.

Tongues can be cut in several ways on the router table. With a bit height already established for the groove cut, resetting the fence distance is all that's required to cut the tongue. Cut one side of the tongue, flip the board face for face and trim the other side to fit. The only drawback to this method is the potential for tearout when cutting against the grain. A simple way to prevent it is to use what's called a climb cut.

As everyone knows who's read this far, cuts on the router table are made from right to left, or into the rotation of the bit. What happens when you cut backwards (from left to right)? The first thing is that the bit tries to push the work away from the fence or climb out of the cut. You have to fight it to keep it in close. The other result is that the bit wants to take and shoot the board past it. This can be dangerous. But with a large board and a firm grip, there is yet another result from this perilous cut. As the edge of the board gets trimmed, all its fibers are always backed up, and the cut will be clean.

A climb cut is made to score the edge of a piece (see the photo above right). Don't try to push the board in tight to the fence. I let the bit push the work away from itself, but I hold the piece in just tight enough to score the edge. I also use one end of the board always against the fence, much like a starting pin. After the climb cut is made, the board is fed in the

A climb cut made first on the edge of a board helps prevent tearout on its face.

proper direction from right to left and tight up against the fence to clean up the entire cut.

Router table with a slotting cutter

Slotting cutters can be used to create both parts of a tongue-and-groove joint. Most of them cut a pretty deep groove, which generally creates more trouble than it solves for this joint. I use them set up within the recess of a fence to limit the depth of cut. The boards are laid flat on the router table for the cuts.

To produce a tongue to match a groove cut, only the bit height needs changing. Tongue-and-groove bit sets are also sold, but weigh their cost against their potential use. The same must be said for rail and stile cutters. They do provide the ability to cut molding patterns into frames, but they don't offer much in the way of a gluing surface because of their limited depth of cut.

FRAME WORK

In frame-and-panel work, the panel generally fits into a continuous cut made in the frame. This cut, a groove or a rabbet, may be made on the individual frame members or with the entire frame dry-assembled.

Routing grooves in frame members

Grooving a frame to hold a panel in place needs to occur before glue-up so the panel can be inserted. One method of grooving is to take each individual member of a frame and pass it by a straight bit on the router table. Unless a stub tenon lines up with this groove cut in the frame, the cuts will have to be stopped cuts. You can clamp stops to the fence to limit the travel of the board to the bit (see p. 51).

The advantage to this method of grooving is that each piece is easily handled, and the stopped cuts are quickly cleaned up. However, if the frame twists a bit as it goes together, grooves that are cut perfectly straight may not line up. In that case the panel has to be taken down in size in order to fit into the misaligned grooves.

Routing grooves in dry-clamped frames

Frames that fit tightly can be put together dry and then grooved for a panel using a slotting cutter. This can be done to small frames on the router table. Larger frames are cut with a router held topside (see the top photo on the facing page). If the frame has flush stiles and rails, clean up the face of the frame before grooving it. Then the router base won't hit any uneven surfaces as it moves around the frame. Offset faces in a frame require spacer blocks laid down onto the lower frame parts for the router base to ride on.

Start the router off the surface of the frame, get it up to speed, and then place it flat on the frame. Bring the cutter and bearing straight into contact with the wood. There can be absolutely no tipping of the router or you'll find yourself doing patch work. First run a climb cut moving counterclockwise inside the frame to prevent tearout. Then make the full cut to depth in the proper clockwise direction inside the frame. For grooves that are wider than your slotting-cutter width, take two passes to get to size.

The corners of this slotting cut will be left quite round, and so the frame must be disassembled and the corners chopped square before fitting a panel to it.

Rabbeting a frame

Not all frames get grooved for a panel; some are rabbeted on their front or back face instead. This is the standard method for any panel that may need replacing or one used with decorative moldings.

To groove for a panel topside with a slotting cutter, first raise the frame off the bench top with spacer blocks so the bearing of the cutter won't mar the surface. Clamp the frame securely in place, and make sure it's well supported and will not flex with the weight of the router on it.

In most cases, there are stopped and through cuts that need to be made in the frame. A rabbeting bit can do the job with the frame glued together and with no concerns about stopping a cut.

Rabbeting bits are available now with a variety of bearing sizes (see the photo at right). A standard rabbeting bit comes with a cutter and bearing that allow a ³⁄₈-in. depth of cut into the work. With the rabbet sets now being made, rabbets from as small as ⅛ in. up to a full ½ in. can be made by simply changing the bearing on the cutter.

Some rabbeting bits come with a variety of removable bearings. Their different diameters yield various depths of cut.

RABBETING A
MIRROR FRAME

Here's how I cut a rabbet to hold the mirror in the half-lapped mirror frame in the sidebar on pp. 126-127. After the frame was completely glued up and shaped, I cleaned up both of its faces using a hand plane and a cabinet scraper. Then I used a ⅜-in. rabbeting bit in a hand-held router. (I should caution you again to make sure, before turning on the router, that your bit can make a full rotation within the router base. A bit that cannot spin freely will tear right into your base and either ruin itself or your motor.)

A frame that is rabbeted with a router held topside must usually be raised off a bench surface to protect the bench

from the router-bit screw that holds in the bearing, which projects down so far that it can damage a benchtop. I put spacers underneath my frame and clamp the frame firmly down to my bench. Putting it between a vise and some bench dogs is even better because there are no clamps in the way.

The rabbeting bit has a tendency to cause some tearout as it slices through unsupported wood fibers at the edge of the work, so I always use a climb cut first to score the edge of the frame (see the photo below). The cut is very light along the edge, just enough to create a small rabbet. Then the normal pass is made to full depth.

When the rabbeting bit starts cutting at the corner, you'll hear the motor bog down a little, but that's only because the bit is now taking out a whole lot more wood. Don't panic, and always keep the router base flat on the surface of the frame. Make a series of passes to get down to depth. In this frame the rabbet is ½ in. deep.

One of the things you must remember about topside routing is that your hand pressure transfers as you rout around a workpiece. First one hand, then the other, will be over the frame as you move around its edges. Keep better pressure down with the hand that is right over the frame. Lighten your grip with the

hand that moves over the panel area. This will keep the base referencing properly off the frame.

Don't ever let the router dip into the work. But if the tool does dip a little or some tearout occurs, take one more cleanup pass at a 1/32-in. to 1/16-in. deeper setting, and concentrate on keeping that router base flat.

Once the rabbet is cut, you'll notice that the corners of the rabbet are round. They will need to be squared up with a chisel. This saves you a buck or two at the glass place so they won't have to round the edges of the mirror.

To rabbet the mirror frame, make a climb cut just along the edge, first moving from right to left (counterclockwise), and make sure you keep a careful grip on the router. Once this scoring cut is made, reverse your feed direction and cut from left to right (clockwise) at the full depth of cut.

PANEL WORK

Panels may be flat or raised and fit into a groove or rabbet in the frame. If a panel is flat, its entire thickness can simply be fit to the groove in the frame. That's how I made the nightstand shown in the photo on p. 122. If a panel is to be raised, you will have to make some kind of edge cut around the panel. A decision about the kind of raised panel must, of course, come before you select a bit.

Raising a panel on the router table

You can raise panels with the workpiece placed either horizontally or vertically on the router table.

Horizontal panel-raising bit There are a number of horizontal panel-raising bits available for use in the router table. These bearing-mounted bits have very large diameters and require a motor with at least 2 hp with soft start and a variable-speed adjustment on it to lower the speed of the bit.

Vertical panel-raising bit Vertical panel-raising bits can do the same job as horizontal cutters without all that bit edge flying around trying to keep pace with the motor. There are no bearings on these bits so it's important to have a tall fence for the router table (see the photo above right). Make the cross-grain cuts first, followed by the long-grain cuts, to eliminate any tearout problems. If the pattern allows, trim off any waste on the table saw first. If it doesn't, as with any large profile bit, take a couple of passes to get down to the full depth of cut. This method doesn't make my fingers as nervous as one big hogging pass.

A vertical panel-raising bit is used with a tall auxiliary fence to support the panel. Cut a recess in the fence to house the bit.

Slotting cutter A method that I often use for my raised-panel work uses the same slotting cutter that cuts the grooves in the frame. The groove is made into the frame at a $\frac{7}{16}$-in. depth (see the drawing on p. 142), which is as deep as the bearing will allow the bit to enter. I make up my panel at the thickness I want and then cut its length and width to fit my frame, less just a little. I take the inside dimensions of my frame plus the full-groove depth and then subtract $\frac{1}{8}$ in. on all sides.

Then, with the slotting cutter set up in the router table, I cut a rabbet on my panel at that same $\frac{7}{16}$-in. depth of cut. When the panel fits into the frame, there will

FRAME AND RAISED PANEL

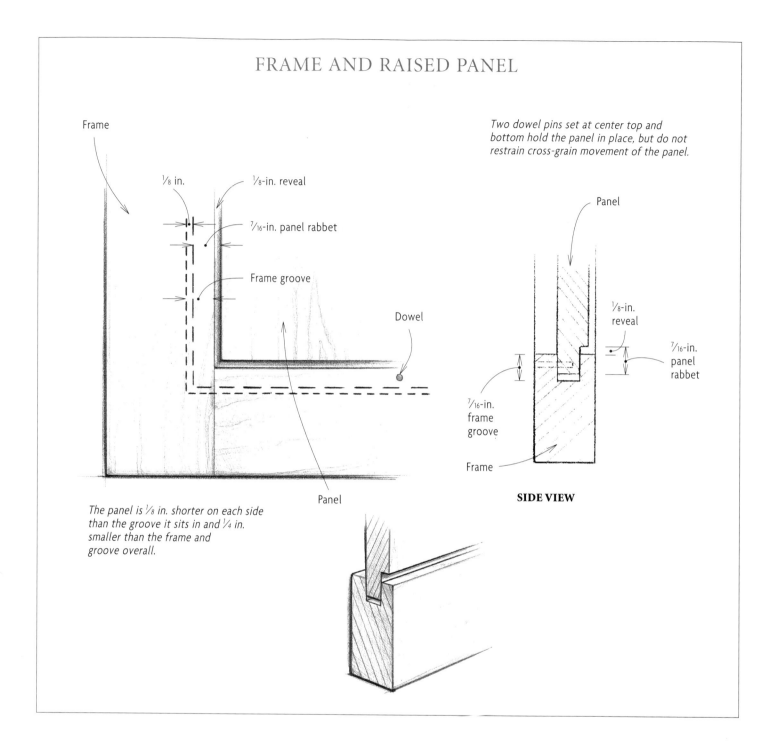

Frame

⅛ in.

⅛-in. reveal

⁷⁄₁₆-in. panel rabbet

Frame groove

Dowel

Two dowel pins set at center top and bottom hold the panel in place, but do not restrain cross-grain movement of the panel.

Panel

⅛-in. reveal

⁷⁄₁₆-in. panel rabbet

⁷⁄₁₆-in. frame groove

Frame

SIDE VIEW

The panel is ⅛ in. shorter on each side than the groove it sits in and ¼ in. smaller than the frame and groove overall.

Panel

be a ⅛-in. reveal or groove all the way around the panel, giving a nice shadow line to the design (see the photo on the facing page).

The panel cuts have to be made cross grain first and then along the long grain to remove the tearout

that will happen cross grain. Don't worry about using a backer board. As long as you follow every cross-grain cut with a long-grain one, you will remove the tearout.

Clean up the rabbet cut with a scraper and some sandpaper

A slotting cutter cut the rabbets on these highly figured panels and also cut the slots in the frame.

wrapped around a square block of wood. Don't worry about cross-grain scratches. Start with 120 grit and work up to 220. The scratches in that reveal will be too small and hard to see to worry about. When the rabbets are sanded, fit the panel to the frame by planing or scraping the back of the panel to fit. I use my smoothing plane and leave the hand-planing marks on the panel back.

Edge-jointing large panels

There are situations in the shop when your jointer just isn't big enough to handle a large piece of wood. When edge work needs to be done on large boards, tabletops, or panels, a router can be used to joint edges for gluing.

There are two methods of using the router as a jointer. One is topside with a straight bit, using a two-part fence to direct the cut. The other is using an offset fence on the router table. Use the bit with the largest diameter you have to give a cleaner cut.

Topside with a two-part fence When I built the library table shown in the photo on p. 56, I edge-jointed the top with my router once I had the center section glued together. This panel was 40 in. wide and too large for my 6-in. jointer, but I still had two boards to add on.

To make the cuts I used a straight bit in my fixed-base router, along with a sub-base milled with a straight edge. The fence is made up of a piece of ¾-in. MDF glued to a piece of ¼-in. MDF. The fence pieces are offset exactly the same as the distance from the straight bit to the sub-base edge. This is simply done after the pieces are glued together by making a cut along the ¼-in. MDF with the sub-base and bit you'll be using for the cut. This way I could set the fence exactly where I wanted the cut to be and use my largest straight bit (see the photo below). If you use a flush-trimming bit instead to make the cut, just make up a single straight fence and clamp it in place.

Take your time with this cut. Just as on the jointer, the slower the feed rate, the more cuts or revolutions per foot. With a fast feed rate, the cut won't be as flat.

Router-table setup The router table can also be set up to make a jointer-like cut. Set up a fence with a recess for a large straight bit to hide in. Then, if you have a split

fence, move the outfeed side of it to be exactly in line with the cutting edge of the bit. Spin the bit by hand so that one flute is at a point farthest from the fence, and line up the outfeed fence with this. Adjust the infeed side to allow a pass into the bit that's about ¹⁄₆₄ in. to ¹⁄₃₂ in. deep.

If your fence can't adjust in this fashion, don't despair. Just get a piece of thin veneer or laminate that's flat, wide, and of a consistent thickness. Attach the scrap to the outfeed table with double-stick tape, and expose the bit to make a cut exactly as thick as the veneer or laminate (see the drawing on the facing page). The work will first index against the infeed side of the

To edge-joint a large panel, use a straight bit, a sub-base with a straight edge, and a two-part fence. Place the fence exactly where the cut should go, and move the router slowly along the edge of the workpiece.

fence. Then it will be cut by the bit taking a pass as deep as the scrap is thick. As the piece moves past the bit, it contacts the scrap behind the bit and will be fully supported. Start with your hand pressure on the infeed side to begin the cut and finish with this pressure on the outfeed side of the bit.

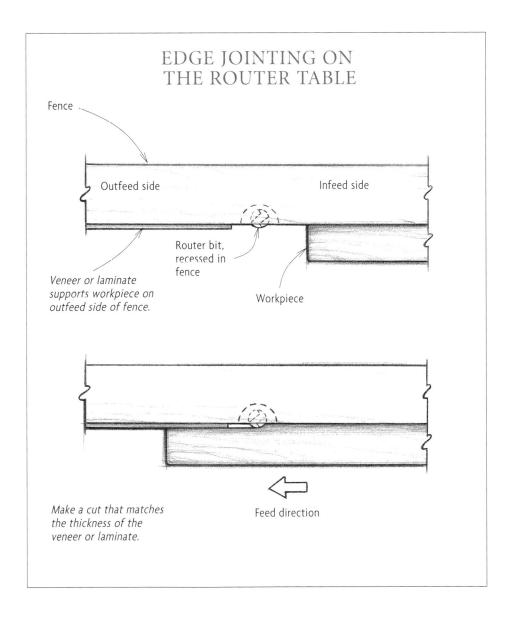

EDGE JOINTING ON THE ROUTER TABLE

Fence

Outfeed side

Infeed side

Router bit, recessed in fence

Veneer or laminate supports workpiece on outfeed side of fence.

Workpiece

Make a cut that matches the thickness of the veneer or laminate.

Feed direction

CHAPTER NINE

STOOL AND TABLE CONSTRUCTION

This dining table by the author is a good example of leg-and-apron construction. Photo by Harold Wood.

I built my first table out of red oak. The mortises were chopped out by hand, and I fit the tenons flawlessly, working that oak with care. I took that table to my first crafts fair—this was 1975—and no one, to my amazement, noticed the beauty of these joints (of course, they were hidden). We woodworkers sometimes like people to take notice of our efforts. Joinery, however, is one aspect of woodworking that is best appreciated as time goes by, long after the tools have been put away.

There are two broad systems for building tables, stools, and similar pieces of furniture (see the drawing at right). One approach uses leg-and-rail construction. This is very similar to frame-and-panel construction, but there are no panels within the frames. The frames are connected together at their corners by sharing a common stile, or leg.

The other approach to building stools and tables uses trestles or pedestals to hold up the top surface. Trestles can be connected with rails or stretchers. The joinery employed in putting these forms together is identical to that used for either frame or carcase construction. Large, wide panels functioning as legs can also be joined at their corners to benchtops or tabletops using dovetails, finger joints, or other carcase joints.

Each of these construction methods for stools and tables borrows something from carcase or frame-and-panel construction and adapts it for use. Those methods of joinery are covered in Chapter 7 and Chapter 8, respectively. This

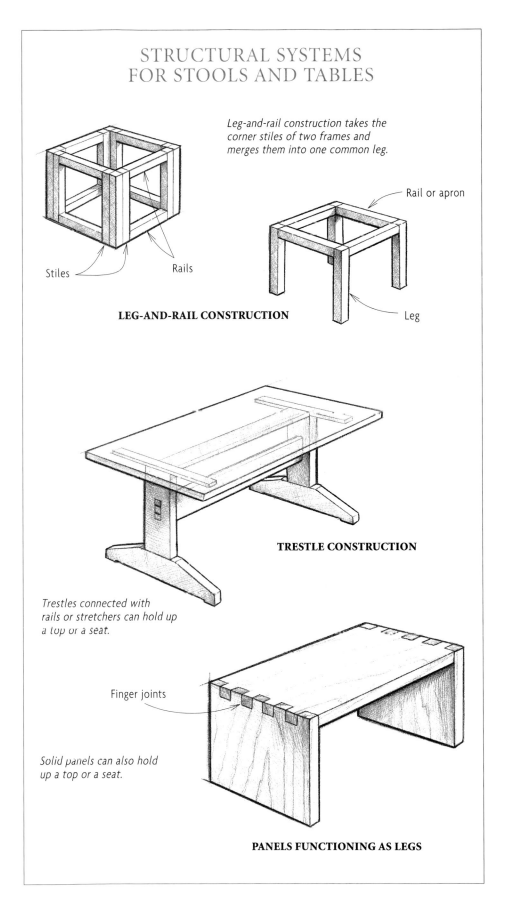

STRUCTURAL SYSTEMS FOR STOOLS AND TABLES

Leg-and-rail construction takes the corner stiles of two frames and merges them into one common leg.

Rail or apron

Stiles

Rails

Leg

LEG-AND-RAIL CONSTRUCTION

TRESTLE CONSTRUCTION

Trestles connected with rails or stretchers can hold up a top or a seat.

Finger joints

Solid panels can also hold up a top or a seat.

PANELS FUNCTIONING AS LEGS

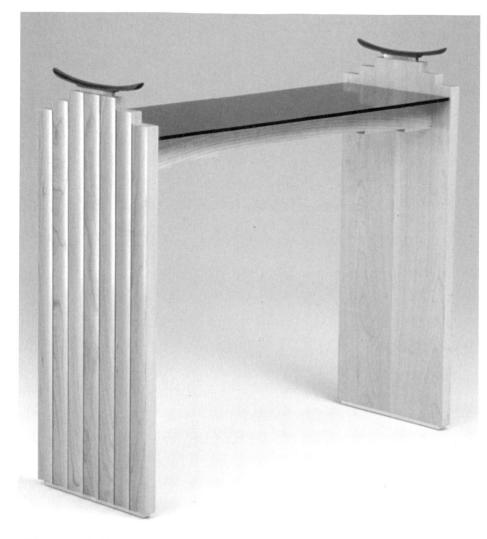

This cherry hall table by the author uses trestle construction. Photo by Jim Piper.

variations of this joint that they have been given their separate names, like finger joints or dovetails. Webster's even defines a dovetail as an angled mortise and tenon. But the simple mortise-and-tenon joint also has many versions (see the drawing on the facing page). Consider what makes up the joint and you'll see why.

Mortises

A mortise is generally cut at right angles into a board. The side walls remain parallel throughout while the ends of the mortise can be parallel or angled out to allow a wedge to spread the tenon. The mortise can go all the way through a board or stop short. It can be centered in the stock or offset to one side or another. The mortise ends can be left round or chopped square.

Tenons

Tenons (see the drawing on p. 150) are cut with cheeks and shoulders. The cheeks expose long-grain glue surfaces, and the shoulders are cuts across the grain that provide support and resistance to movement. Two cheeks and two shoulders are usually cut, but shoulders can be cut into all sides of a tenoned board. Cheeks and shoulders are generally cut square, but either or both can be angled for some constructions, such as seating pieces.

Haunches are used when tenon cheeks are not cut to the full width of a piece, but there remains a need to prevent twisting, as in a table rail. In very wide boards, haunches can also be placed in the middle of the joint and on its ends

chapter will focus on the principal method used for stool and table construction: the mortise and tenon.

MORTISE-AND-TENON BASICS

Except for the butt joint and its cousin the miter, every other joint is really some form of the mortise and tenon. Simply stated, a hole (the mortise) is cut into one board to accept a tongue (the tenon) cut in another. There are so many

BASIC MORTISE-AND-TENON JOINTS

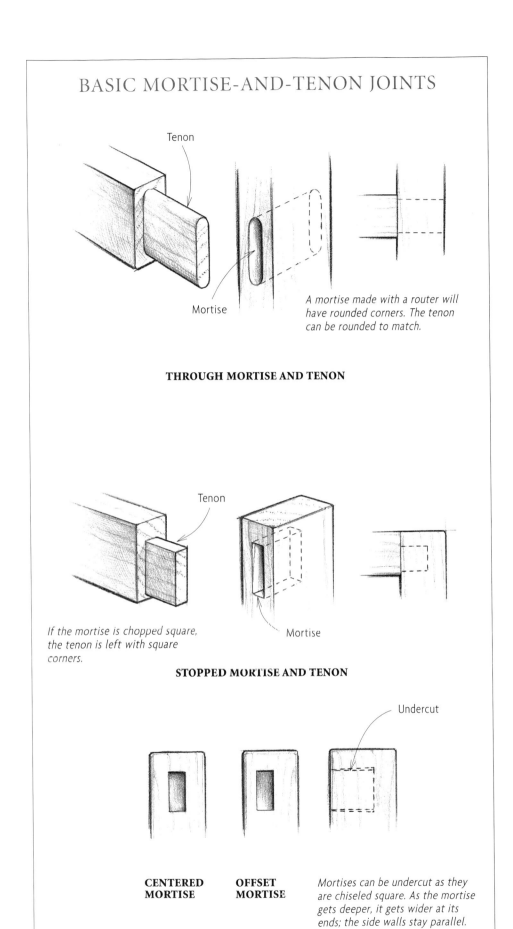

Tenon

Mortise

A mortise made with a router will have rounded corners. The tenon can be rounded to match.

THROUGH MORTISE AND TENON

Tenon

Mortise

If the mortise is chopped square, the tenon is left with square corners.

STOPPED MORTISE AND TENON

Undercut

CENTERED MORTISE

OFFSET MORTISE

Mortises can be undercut as they are chiseled square. As the mortise gets deeper, it gets wider at its ends; the side walls stay parallel.

Use a drift made out of a thin piece of scrap to check the side walls of the mortise. If it hangs up anywhere, that's where the joint is too tight. If the mortise has been cut well, the drift should fit consistently in all parts of the joint.

In ¾-in. stock, I use a ¼-in. bit to cut the mortise and no larger. But in ⅞-in. stock I move up to a ⅜-in. bit and rout with that. The gluing surface remains identical, but I think the compression strength is greater for the tenon. In 1-in. stock I use ½-in. tenons. After all, tenons support the load, so I don't undersize them. I also leave a ¹⁄₁₆-in. shoulder at the bottom of the tenon to cover up any defects around the mortise opening.

Since the ends of the mortise will be cut round by the router bit, I either chop them square or leave them round and round the tenon. If I chop them, I always undercut, making the mortise wider as it gets deeper. The mortise cannot get narrower or fitting the tenon would be really tough, and trying to chop the ends perfectly square isn't really necessary either. These end-grain surfaces are not important for gluing, so under-cutting them doesn't affect the joint's strength. Check this chopping cut with a combination square or a small rule to be sure the mortise ends are straight or angle slightly out.

The mortise side-wall corners need paring as well when this chopping cut is made. These walls must remain parallel. A simple way to check your work is to use what I call a drift (see the photo above left). It's a piece of scrap cut on the bandsaw and hand-planed to fit the width of the mortise. Since the drift is checking only a very thin section of the mortise at a time, it's easy to figure out where you need to pare.

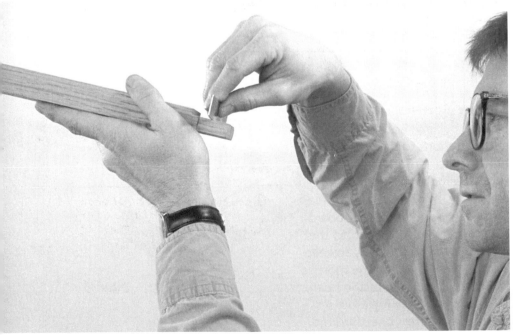

If the mortise is right and ready, then check the tenon for the high spots with a straightedge. If the fit is still tight, hold the tenon up to the light. The high spots will be shiny from rubbing against the mortise.

Fitting mortise-and-tenon joints

On your best days in the shop, fitting mortise-and-tenon joints can be a simple and pleasing task. The tenons slide home with little fuss and plenty of whistling and humming on your part. Then there are the days when fitting tenons gives you the fits. You check the tenon in a mortise, and it won't go. You pare it carefully, and it enters halfway and then stops. No amount of pushing gets it in any farther. And then when one more shaving is taken, the tenon is suddenly too loose and just about falls out of the mortise. Ah, the joys of woodworking.

The rules for fitting mortise and tenons are simple but inexorable. You have to pay attention to certain things or your frustration level will rise to new heights. After the mortise is cut and chopped perfectly, work on trimming and fitting the tenon. If a tenon doesn't quite seat in a mortise, find out where the trouble area is by holding the tenoned piece up to a light source (see the bottom photo on the facing page). If you see any shiny spots on the cheek of the tenon, that's where it's rubbing and is tight. Mark these spots with a pencil and pare them off with a rabbet plane or chisel. Don't sand or file a joint to remove material, because this will round over the cheeks.

The fit you're looking for is somewhere between a tight piston fit and one so loose the joint falls apart on its own. A good-fitting joint requires just a little bit of persuasion or pressure to put together. Too tight a fit can split the wood around a mortise, especially with glue inside the

CHECKING FOR A FLUSH FIT

Fitting a tenon to a mortise doesn't require the patience of a saint, just some straightforward techniques. But there is only a small difference between just right and too tight. Other surfaces need careful alignment as well. If a leg and rail are designed to have flush outer faces, here is a simple method for checking their final fit.

With one tenon cheek cut, place the tenoned rail up to the mortise, the way it should go in. Then turn the rail around and set the tenon shoulder down on the surface of the mortise (see the drawing below). The face of the rail should line up with the mortise if the tenon cheek is cut properly. If it doesn't quite meet up, then there's more to trim off the tenon. If it extends beyond the mortise, too much has already been removed, and it's time to pull out some little tenon scraps (see p. 165) for some repair work.

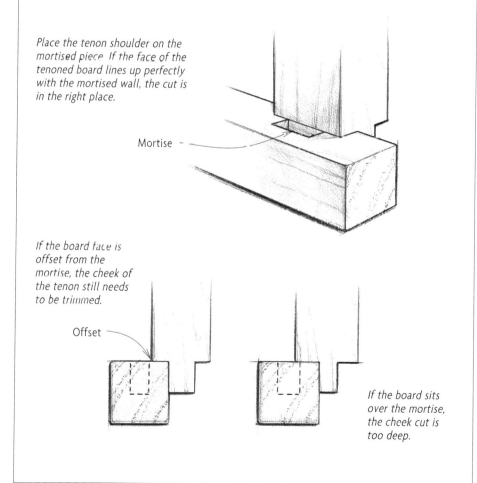

Place the tenon shoulder on the mortised piece. If the face of the tenoned board lines up perfectly with the mortised wall, the cut is in the right place.

Mortise

If the board face is offset from the mortise, the cheek of the tenon still needs to be trimmed.

Offset

If the board sits over the mortise, the cheek cut is too deep.

joint. Don't use a hammer either, which tends to mash ends and force joints together. On the other hand, a loose-fitting joint will need a stronger, more gap-filling adhesive than yellow glue.

If a tight joint won't come apart, you can use a dead-blow hammer to separate the members. Or clamp the workpiece down to your bench and then pull the pieces apart. Be sure you pull the tenoned piece straight out of the mortise so the mortised piece won't be damaged.

One fault that beginners often make when fitting tenons is to make all four sides fit snugly into the mortise. Remember that the cheeks are the most important part of the joint, so that's the place to concentrate on a close fit. The ends of the tenon will be next to the end grain inside the mortise, so don't bother placing glue there. The mortise ends can be undercut (see the drawing on p. 149) and the tenon width kept undersize.

A tenon should always be about $\frac{1}{32}$ in. to $\frac{1}{16}$ in. shorter than the full depth of the mortise to allow room for excess glue inside the joint and to ensure that the tenon shoulders will pull all the way home. Tenons should fit tightly to the leg or stile, giving a clean appearance and providing mechanical resistance for the joint.

STOPPED MORTISES

Mortises that aren't cut all the way through a board are called stopped mortises. The depth can vary, depending on the strength required. A mortise can be as short as $\frac{1}{2}$ in. or as deep as 3 in. My own rule of thumb is that if I have to cut deeply but not through a board, I cut only to within $\frac{1}{4}$ in. of its far edge and no closer.

Stopped mortises can be cut in several ways with the router. The method you choose may depend as much on the type of router you have as on the results you want to achieve. The depth of a mortise can depend on what length bit you have available. If you need an especially deep mortise, use a longer bit with a shank size that matches or is smaller than the bit diameter. That way it can penetrate into the mortise without affecting the mortise size. If your bit is still too short, don't resort to pulling the bit farther out of the collet to gain extra length. You only run the risk of the bit pulling loose from the collet. Buy a longer bit, or redesign the joint.

Plunge router with fence
The best way to cut mortises topside is with a plunge router. A fixed-base router can be used, but I don't recommend it. With the bit sticking out and the base rarely in contact with the workpiece, it's a difficult cut to make accurately. This mortising work is exactly what the plunge router was made for.

The beauty of the plunge router is that after each pass the bit is plunged into the cut just a little deeper to make the next pass. There is no resetting of the bit depth or turning off of the router.

You can plunge to depth and lock the bit in place with the locking handle, but I prefer just to move the router down to the next depth I want, hold it there, and make my cut.

The full depth of cut can be set either on the router's depth gauge or against the bit protruding out of the base (remember to cut the mortise about $\frac{1}{32}$ in. deeper than the final tenon length). The placement of the mortise walls is set by using the straight fence mounted to the plunge router. With the router unplugged, rotate the bit by hand until one cutting edge is closest to the fence, then measure from this edge to the fence.

I try to do most of my cutting moving left to right so the bit pulls the fence into the work, but you can take a full pass in both directions if you're careful to hold the router and fence in tight to the work.

Is it only a steady hand that brings the router just up to the ends of the mortise and no farther, or can help be found? One simple trick is to use pencil-marked stops against the outside of the router base. These pencil marks are easier to see than the ends of the mortise through all those chips flying about. Mark out the position of the router base when the bit is at the ends of the mortise cut (see the photo above right). Then move the router just up to this pencil mark each time when cutting.

A more precise method for indexing the length of cut is to clamp stops directly to the workpiece. Line up the bit again on the edge of the mortise and then clamp a stop right up against the router base—repeat at the other

With the bit in position over the mortise, place a pencil mark right against the router base to serve as a warning. Move just up to this mark on each pass to index the cut.

end of the mortise. These stops limit the travel of the router, and hence the length of the mortise.

Plunge router with template

All this measuring and marking can grow tiresome if you're mortising legs for a set of eight chairs. This is exactly why templates are so handy. By creating a pattern that can be used over and over, your setup time is dramatically decreased. Making the template is time saved if it will be used for another run of the design or a different project that requires mortises of the same size.

MORTISING TEMPLATE

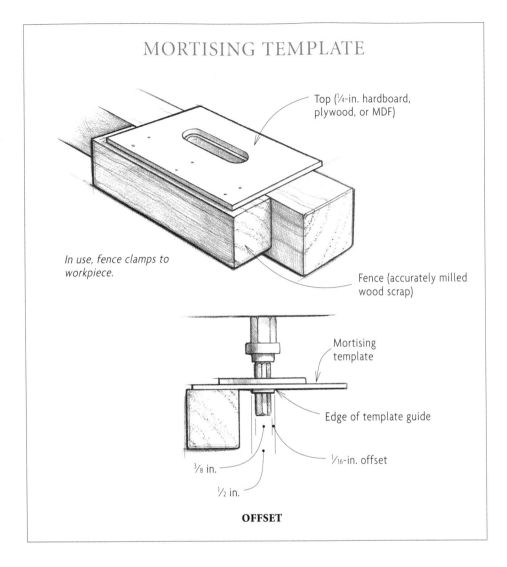

Top (¼-in. hardboard, plywood, or MDF)

In use, fence clamps to workpiece.

Fence (accurately milled wood scrap)

Mortising template

Edge of template guide

¹⁄₁₆-in. offset

³⁄₈ in.

½ in.

OFFSET

The templates I use are made up of two parts: a top and a fence (see the drawing at left). The top portion of the template has a slot cut in it for a template guide to ride in. An accurately milled fence locates the mortise slot the desired distance back from the edge of the workpiece. The template is clamped onto the workpiece to position the slot over the mortise position.

The template is attached to the fence with nails or screws, but it's set back from the edge of it a little. The size of the slot is determined by the size of the mortise and the template guide. Since a template guide is used, the slot must be made larger all around than the final size of the mortise. How much larger depends on the difference in diameters of the bit and the template guide. Let's say you want to use a ³⁄₈-in.-dia. bit and a template guide with an outside diameter of ½ in. The difference in diameters is ⅛ in., so the offset to each side is half that, or ¹⁄₁₆ in., as shown in the detail drawing at left. So the slot in the template will be ½ in. wide to fit the guide, and ⅛ in. longer overall than the final length of the mortise. For a mortise wider than the bit being used, the slot needs to be correspondingly oversize in width.

Now mark the slot to be cut on the underside of the template (see the photo at left). The slot's placement along the length of the template is not critical. Then set up your router table to make the cut. I fully expect there to be some fussing time to this. Just remember that each mortise will take only minutes to cut if you set up the template accurately.

Mark the length and width of the slot and the distance of it from the fence on the underside of the mortising template.

If the gods are smiling, you will have a bit that's exactly the right size to cut the slot needed. Since this isn't always the case, you may have to make two passes to get to full width, using a shim of the proper thickness between the router-table fence and the template fence to make the first cut (see the top photo at right). (Always make a practice cut before committing your template.)

Make the first pass with the shim in place, starting the cut roughly in the center of the slot, then remove the shim for the second pass. Make sure the template guide runs smoothly in the slot. If it sticks, waxing the slot will help. When the slot is cut, file its ends flat.

Since the template has no stop on it to register off the end of a board, it can be placed to locate a mortise cut anywhere along the length of a board. But this also means that one mortise end must be marked out for each cut, with the template slot offset the proper distance (in our example, $\frac{1}{16}$ in.) to this mark. Clamp the template onto the work, making sure it rests flat on the workpiece (see the bottom photo at right). To place the mortise closer to a board's face, use shims between the fence and the workpiece.

Rout the mortise, moving the router end to end inside the template. As you rout, the debris from the cut will probably pile up in the template slot, obstructing the template guide. It's convenient to have an air hose nearby (or a good set of lungs) to blow away these chips. An alternative method of cutting is to plunge to depth at each end of the mortise first, then

Cut the mortising slot on the router table, with the template upside-down. Use a shim to push the template out away from the fence for the first cut. Note the practice cut at one end, which was used to check the fence distance before plunging into the template.

A mortising template set up to cut. The board is clamped securely to the bench. Position the C-clamps where they won't interfere with the router base.

MAKING A MORTISING JIG

The key to building this jig is having accurately milled stock put together at exact right angles. This seemingly formidable task is not so difficult to achieve if a few simple details are looked to. First of all, use high-quality sheet goods to eliminate any warping or cupping. Use ¾-in. MDF or plywood for this jig, but check to see that it's flat and consistent in thickness.

This jig (see the top photo on the facing page) is really just a tall U-shaped box. Most table or chair legs—the parts you are most likely to mortise—are under 3 in. thick, but to allow for any situation, make the jig 4 in. to 6 in.

high and 18 in. to 24 in. long. Mill up accurate spacer blocks to raise thin stock higher up in the jig. The bottom of the jig should be wide enough to allow even curved pieces within its walls, but it can't be wider than the plunge-router base that will ride on top of the jig. For most work, a width of 3½ in. should be adequate.

To make it simpler to glue up and to ensure accurate placement of the side walls, cut a ⅜-in. by ¾-in. rabbet into their bottom edges. The side walls will then line up at exactly the same height relative to one another and to the jig bottom. The sides are screwed to the bottom.

On my jig I don't use adjustable stops because they can shift if knocked around too much. But they could be set onto the top or the sides of the side walls and slotted for a bolt or screw to fit through. The screw or bolt will mate into a barrel nut, but make sure the setup can be securely tightened.

Dedicated mortising jigs can also be made that have stops for length of travel and board placement permanently attached. Accurately sized mortises can be cut with minimum setup time this way. One such jig, shown in the photo below, cuts angled mortises for a stool design.

This dedicated mortising jig cuts angled mortises for a stool project. It has stops permanently fixed in place. The straight fence mounted with an auxiliary fence runs between the stops to cut the proper sized mortises each time.

clean out the waste between these holes. This will help with the chip removal. However, be careful of excessive burning at the ends of the mortise unless you have a bit with center-cutting capacity.

Plunge router with mortising jig

Templates are useful, but limited because they are specific for a job. What about a fixture that is more universal? A mortising jig used with the plunge router and a straight fence provides this flexibility. This is a jig you can build yourself (see the sidebar on the facing page). It has room for stops to locate boards and index cuts, and a stable surface for the router and fence to ride along (see the top photo at right). All the workpieces are placed in the jig in the same spot for accurate results. Mark out one mortise to be cut and clamp that board in the jig. A stop placed at the end of the board indexes all subsequent cuts.

The router fence must be set to put the cut in the proper spot on the board. Lay the router on the jig with the fence tight up against the side wall. Move the router until the bit lines up with the mortise mark, and then lock the fence in place. The mortise length is set with stops that limit the travel of the router fence. Rotate the bit by hand to line up a cutting edge on a mortise end wall, and clamp stops onto the outside wall of the jig (see the bottom photo at right).

The depth of cut is set relative to the height of the workpiece when it's in the jig. The bit is first zeroed down onto the work. Then the depth gauge is measured and set. If the board is too low in the jig, add spacer blocks underneath it to raise it higher.

This shop-made mortising jig uses clamps to hold the workpiece and to set stops.

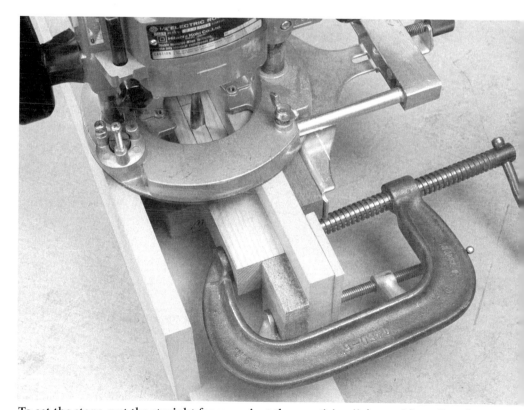

To set the stops, put the straight fence against the mortising jig's outside wall and move the router bit to the mortise. Have the bit rotated so one cutting edge will line up with the side wall of the mortise. Then set the stops on the jig with the bit rotated to line up with the ends of the mortise.

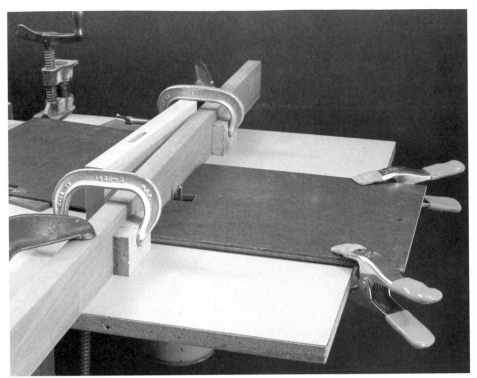

Mortising on the router table with a fixed-base router. Instead of continually resetting bit depth, start with a stack of thin, flat table shims. After each pass, pull out one shim to go to the next bit height. Feed direction is always right to left.

One method on the router table is to set the bit height ⅛ in. deeper for each pass. The fence is set to put the bit the right distance from the edge of the workpiece, and stops are clamped to the fence to limit the length of the mortise. The piece is always fed from right to left against the fence. Remember to hold the work in tight to the fence when raising or lowering it onto the bit. This method is slow, and the router needs to move up and down sweetly and accurately within its base.

But that's just the problem with a fixed-base router. Most of them have difficulty keeping the bit perfectly centered in the base each time it's moved, and any play will cause a series of steps in a mortise.

A method that sidesteps this issue is to set the bit for a full depth of cut and use a series of table shims. Table shims (see the photo at left) are made of ⅛-in. or ¼-in. flat stock, notched to allow the router bit to poke through. Put down enough shims to almost cover the bit and set the fence and stops normally. After each pass, pull one of the shims to expose more bit.

When routing, make your cutting passes moving left to right against the jig so that the router and fence are pulled into the jig. Only you prevent the router from pulling away from the jig and ruining the cut, so it's important to be aware of this risk during the mortising.

Router-table mortises

Mortises can be cut on the router table with a fence and clamped-on stops. Here, a fixed-base router can be used more successfully than topside for mortising. Smaller pieces are best handled this way rather than with jigs and clamps draped all over them. With a dust collector, chips are easily collected as well.

THROUGH MORTISES

Mortises cut through a board to be visible on a face are called through mortises. They require some special attention because of the possibility of tearout of unsupported wood fibers.

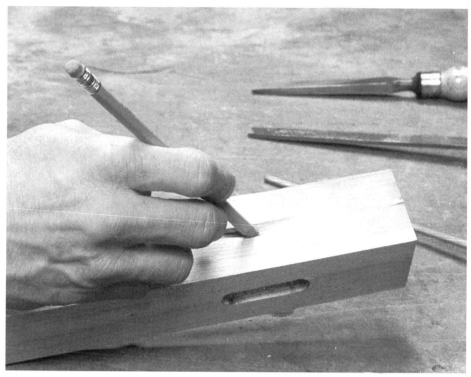

On a through mortise cut stopped just short of the face, a pencil can push through the thin wall that remains. The cut can be cleaned up with a file or chisel.

Router-table mortises

For a through-mortise cut on the router table, set your fences and stops just as for a stopped mortise (see the discussion on the facing page). The obvious difference is that the bit pokes through the wood for the final cut, raising concerns about safety and tearout. It is very important when the bit does come through that your hands are away from it. Also, any tearout will have to be fixed or covered up. A piece of masking tape pressed onto the exit surface of the bit can help to prevent tearout. Another way to prevent tearout is to stop this cut just a whisker short of the face, then clean up that paper-thin remainder of wood with a chisel or smooth-cut file (see the photo above).

Plunge router with mortising jig

Through mortises are easily cut in the mortising jig with a plunge router. The issue is how to protect the jig. Unless it's a dedicated jig with a permanent exit hole, the jig bottom should be covered with a piece of scrap. Alternatively, the height spacer blocks can be sacrificed as backer pieces. Just don't let the scrap move between cuts because then it won't always back up the entire mortise.

Through-wedged mortise and tenon

The drawback to through mortising is the exacting precision required for really fine-looking joints. What has to be one of the marvels of woodworking is how a simple strengthening device like the wedge can also improve the appearance of your joints. A wedge slot is made by cutting into the tenon before it's glued into the mortise. This slot is filled rather neatly with a wedge-shaped piece of wood hammered into place. The wedge forces the tenon into the end walls of the mortise. If the wedge is cut properly, the pressure exerted is enough to close up any little gaps between the tenon and the ends of the mortise.

One item to pay particular attention to when wedging a joint is where the pressure will be exerted (see the drawing on the facing page). If it's placed so that it pushes out along the natural lines of cleavage in a board, the wedge may split the wood. It is always better to orient the wedge so that it will exert pressure against the end grain of a mortise. But even in this situation, there can be danger if the mortise is put too close to the end of a board. Look at each situation with a discerning eye and consider the potential for splitting

the wood before choosing to use wedges. A wedge can apply so much pressure that it can split a tenon at the bottom of the wedge slot (see the photo below left).

For through mortises, I think thin wedges look better than fat ones. To alleviate the splitting problem I drill a 3/16-in.-dia. relief hole two-thirds of the way in from the end of the tenon. In this way the pressure exerted by the wedge gets spread out around a wider area, so the tenon is less likely to split. Then I cut a 3/32-in. slot for the wedge up to the relief hole, using the bandsaw. The wedges are cut on the table saw with a tenoning jig and an angled cut. Their thickness is about 7/32 in., and they are just as wide as the mortise but slightly shorter than the length of the wedge slot. You don't want a wedge bottoming out before it's through doing all its work. These wedges drive nicely and fill the slot sweetly when finally cleaned up.

Knock-down through-wedged mortise and tenon

Knock-down joints are wonderful joinery options when it's necessary to take apart and move a piece of furniture often. Wedges can be placed in through tenons that lock the piece firmly in place (see the drawing on p. 164), yet the joint can be taken apart with some light hammer blows.

The tenon is not glued in the mortise so there's no need for a perfect piston fit. If the joint is going to be knocked down many times, you don't want to have to fight it together or apart. The mortise is cut the same as any through mortise.

A wedged mortise-and-tenon joint cut apart shows the effect that too large a wedge can have on a tenon.

What is important for the strength of this joint is a good set of shoulders on the tenoned rail. These shoulders will mate against the inner face of the mortised column. The wedge also needs to have a good long flat on it or a broad and wide surface. As the wedge is driven into its mortise, it pushes against the far end of this hole, pulling the rail shoulders in tightly into the column. The result is a surprisingly sturdy joint.

Make sure there's enough extra wood left at the end of the tenon after the wedge slot is cut into it. A short-grain situation is created by this slot, and if too little material remains, the end of the rail could break out. How much material to leave depends upon rail, tenon, and wedge size. So err on the side of too much rather than too little wood left behind.

The wedge mortise can go through the tenon horizontally or vertically, depending on how accurate and long a through mortise can be cut through the rail. If it's too wide for a router bit to plunge all the way through, use the drill press to cut the wedge mortise. To cut a through mortise for the wedge, the mortising jig can also be set up, using an angled spacer between the bottom of the rail and the mortising jig to lift the rail. There is a good chance of tearout on the bottom side of the tenon when it is through mortised. Taping on a piece of scrap to the tenon or placing another piece of scrap between the tenon and the jig will prevent most tearout.

WEDGED TENONS

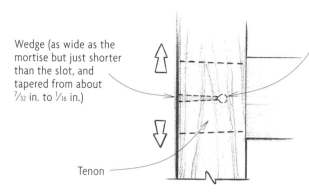

Wedge (as wide as the mortise but just shorter than the slot, and tapered from about 7/32 in. to 1/16 in.)

Relief hole drilled at about two-thirds of the tenon length, at the end of the slot

Tenon

BASIC WEDGED TENON

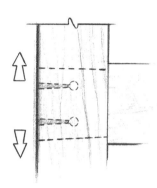

DOUBLE-WEDGED TENON

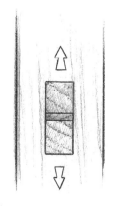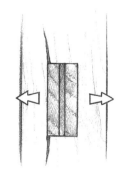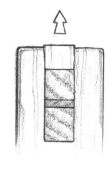

A correctly oriented wedge exerts pressure against the end grain of the mortised board.

This wedge exerts pressure against the side grain, causing a split.

A wedge placed too close to the end of a board can cause the short grain to shear.

WEDGE ORIENTATION AND SPLITTING

KNOCK-DOWN THROUGH
WEDGED MORTISE AND TENON

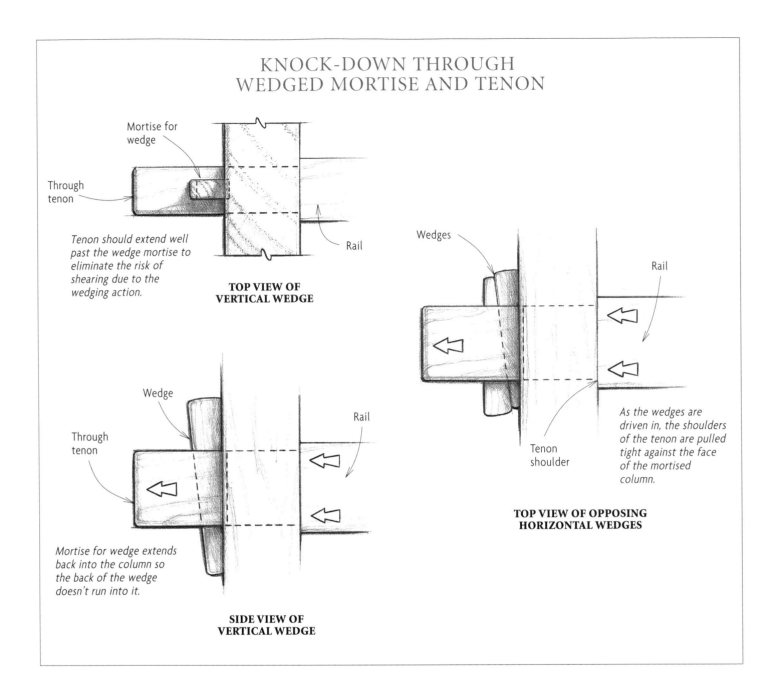

Mortise for wedge

Through tenon

Tenon should extend well past the wedge mortise to eliminate the risk of shearing due to the wedging action.

Rail

**TOP VIEW OF
VERTICAL WEDGE**

Wedge

Through tenon

Rail

Mortise for wedge extends back into the column so the back of the wedge doesn't run into it.

**SIDE VIEW OF
VERTICAL WEDGE**

Wedges

Rail

Tenon shoulder

As the wedges are driven in, the shoulders of the tenon are pulled tight against the face of the mortised column.

**TOP VIEW OF OPPOSING
HORIZONTAL WEDGES**

To provide good wedging action, the wedge should have a slope of about 1:8. Too steep an angle will make the wedge hard to get in at all. It will tend to just bounce out of its slot. Too shallow an angle will not provide enough wedging action for good strength. If you find angled mortises too difficult to cut, you can also use two opposing wedges driven into one large square mortise.

Also make sure that the wedge mortise is wide enough so that the back of the wedge doesn't bottom out on it. Part of the wedge mortise should still be hidden inside the column mortise when the wedge is fully seated, as shown in the drawing.

TENONS

I make my tenons after the mortises, and they can be cut in a variety of ways. I just try to cut and fit them before too much time goes by. The longer joints remain open, the greater their chances of changing size due to moisture loss or gain.

Topside routing

Routing tenons topside allows you to view the workpiece as it's being cut. But topside work like this can cause some problems due to the narrow size of the pieces. Use these methods with caution and weigh their setup times against the other methods available for cutting tenons.

With a straight fence mounted, a series of cuts can be made to establish both the tenon cheeks and the shoulders. Always make sure that the work being routed is clamped firmly to the bench, or that it's secure between a bench dog and the vise. It's best to start cutting at the end of the tenon and gradually work in toward the shoulder. The router will index off the face of each board for its depth of cut, so the stock for a project must be accurately milled. An extra piece to make your practice cuts on is invaluable. Check the depth of cut first before cutting into your good wood.

Before cutting tenons with the router, it helps to get rid of most of the waste first. The bandsaw does a great job of this. Start by marking out one tenon and set the saw fence for a cut that's $\frac{1}{16}$ in. or so larger. With tenons centered in the rail, you can use one fence setup to make all the cheek cuts.

Set up the bandsaw to rough-cut the tenon cheeks, using a stop block to limit the length of the tenon.

Make the first cut, flip the board face for face, and make the second cut. A stop can also be clamped onto the bandsaw fence to limit the length of the cut (see the photo above).

Crosscut the waste pieces off with the saw fence removed. But don't toss these beauties away, because every once in a blue moon, a tenon is cut too loose for its mortise. And with these little scraps saved and reglued onto a tenon that is too loose, you can recut the tenon to fit the mortise.

Index the router pass that cuts the tenon shoulder by setting the straight fence at a distance from the cutting edge of the router bit. The fence will run against the end of each stick, and so these need to be accurately cut. Adding a long

When cutting tenons topside, you can sneak up on a perfect fit by using a doubled-up paper shim between the router base and the board. If the tenon still doesn't fit, unfold the paper and make another pass.

auxiliary fence to the straight fence helps to locate it better to prevent skewing. As with most crosscuts, there is always the danger of tearout. A backer piece can be used to support the cut. Also make sure that the router stays flat on the face of the boards when cutting. Any tipping will mar the tenons. You'll need to support the router carefully when making these or any cleanup passes.

If the tenon is centered in the thickness of board, then only one depth setting is needed. But finding that one setting will take some practice cuts. Cut the cheek of the scrap piece, flip the board over, and take another cut. It will probably take several passes to get down to the right depth to get that perfect fit.

A trick for sneaking up on this fit is to cut one cheek first, and before cutting the other cheek, place a doubled-up sheet of paper between the router base and the board (see the photo above). This will lift your router up just a couple of hairs and prevent too deep a cut. If the tenon is still too large after cutting, unfold the paper shim and take another light pass. Remove the paper shim altogether if more material still has to come off.

Another simple trick is to cut all the cheeks on one side of each piece. Then, working the other side, take a light pass at the corner of one tenon. Try this little section in the mortise to see how close it fits. If it's right, make all the remaining cheek cuts. If not, raise or lower the bit and try again. If it's too deep the tenon won't be ruined because that corner cut is so small.

If your router fence doesn't adjust accurately or if it got lost in the last shop cleanup, there is another method for indexing the shoulder cuts with the router topside. You can clamp a board or fence to a series of rails ganged together and let the router base run against this fence, cutting all the shoulders at one time (see the photo at right).

Begin by clamping all the rails together firmly and accurately, with all their ends lined up. Measure the distance from the cutting edge of the router bit to the edge of the router base, add to it the length of the tenon, and set your fence back this distance from the ends of the sticks. The distance must be the same when you flip over the board to cut the other side of the tenons. The easiest way of indexing this distance is to cut a spacer block out of some scrap that is as wide as the set-in. Then fasten a fence or straightedge against the end of this block (see the drawing at right). This fence will hook over the ends of the sticks to be tenoned and automatically place the spacer block in the proper spot to index the fence.

Router-table cuts

Tenons can also be routed using the router table. The advantage is that narrow boards can be more easily cut there, and longer tenons will present no problems in cutting. When cut topside, there is always the danger of dipping the router into a long tenon. But on the router table, cleanup passes are made with the same kind of good support as a full cut.

Tenons can be gang-cut if the boards are carefully clamped together with their ends lined up. The flat edge of the router sub-base runs against the fence, indexing the cut.

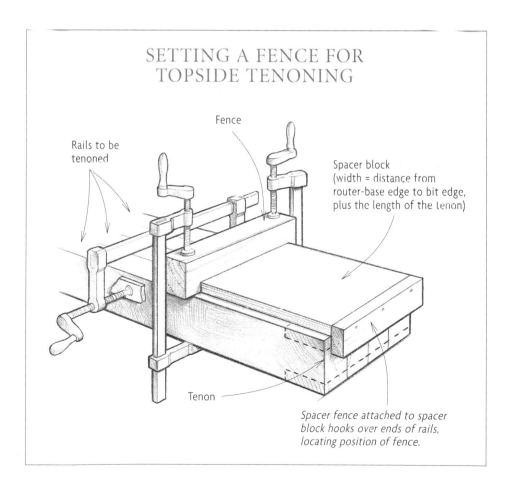

SETTING A FENCE FOR TOPSIDE TENONING

Rails to be tenoned

Fence

Spacer block (width = distance from router-base edge to bit edge, plus the length of the tenon)

Tenon

Spacer fence attached to spacer block hooks over ends of rails, locating position of fence.

Tenons to be cut on the router table can be held together for better support; a backer board prevents tearout at the end of the pass.

down to the final size. After the shoulder cut has been established, the rails can be fed in and out to the fence to clean up any stray sections missed by the first series of passes.

Narrow boards are more easily cut on the router table if they're well supported at their ends. One simple solution is to double or triple the rails being cut, ganging them together in a flat package that will ride against the fence all at once (see the photo at left). This provides a larger bearing surface and greater support. These pieces act as backer boards for one another, preventing tearout. A final backer piece can be added to the package to eliminate tearout on the final rail.

An alternative method, which I use, is simpler. As the bit is cutting the second or third rail in the group, I stop feeding the rails. While still holding the rail that's over the bit tightly to the fence, I pick up the first rail and bring it around to the rear of the group, where it will act as the backer for the last shoulder cut.

Holding the workpiece horizontal
The shoulder cut is established by setting the router fence. If your table has a split fence (see p. 53), place an auxiliary fence over it, or close up the fence as tightly as possible. Rotate the bit by hand so one cutting edge is at a point farthest from the fence, and set the fence distance off that.

Tenon cheeks are cut just like half-laps. With the rail held flat to the router table, a few passes over the bit will be required to clean up the entire tenon. But with the waste cut away, this is relatively quick work. The final pass across the bit will cut the shoulders as well. Make sure that the bit doesn't bog down with too large a cut in this final pass. If need be, trim the shoulder with several passes to get

Holding the workpiece vertical
Tenons can also be cut with the rail held vertically. This method is less tedious than making multiple passes across a horizontally held piece, but as with most techniques, there are trade-offs. You may get some fuzzing at the shoulder cuts, the setups take some time, and the piece must be well supported.

You will need a bit long enough to cut the full length of the tenon. The obvious danger with deep passes and long bits is the amount of flex that can occur in the bit. Try

SUPPORTING A TENON CUT

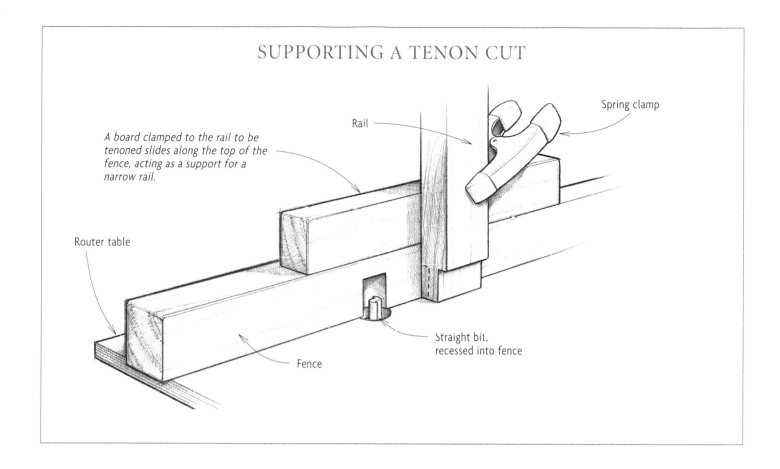

A board clamped to the rail to be tenoned slides along the top of the fence, acting as a support for a narrow rail.

Rail

Spring clamp

Router table

Straight bit, recessed into fence

Fence

to use ½-in.-shank bits for this kind of cut, and take light passes. Another issue is supporting the rail. A table insert should be placed over the bit hole in the router table to prevent the workpiece from diving into it. The fence will have to be notched or adjusted to allow the bit to fit within it as well. A common but dangerous mistake is to set the fence so that the rail will have to pass between it and the bit. If done in the regular feed direction of right to left, the bit then contacts the work and tries to shoot it along. You must always feed work into the rotation of the bit (see pp. 35-37).

A tall auxiliary fence will give greater stability for freehand passes past the bit. This will keep the rails from tipping or leaning as they go

by the bit. Also, a jig like the one used for cutting slot mortises (see the photo on p. 133) can be used to support tenon cuts. Another method is to use a narrow fence that is accurately milled with parallel sides (see the drawing above). A board spring-clamped to the rail being tenoned will ride along the top of the fence, providing additional support for the cut.

You can round a tenon to match the round ends of a routed mortise, or you can chisel the mortise square.

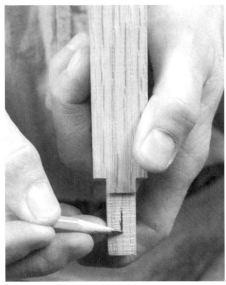

To make a reference line when you're rounding over a tenon end, just locate the line by eye and run your finger against the side of the tenon as a guide.

THROUGH TENONS

Tenons have square edges; routed mortises have rounded edges. If the two are to fit together, you must either make a square hole or round your peg (see the photo above left). In the case of through tenons, this choice will affect the appearance of the joint. The mortise can be chopped square and the tenon cut to fit.

Or do you prefer to round over the tenon on its ends to give the mortise a softer look? Each method has its own merits and challenges.

Chopping a mortise square

To chop a through mortise square, you need precise knife or penciled lines that mark the limits of the mortise on both sides of the board. Chopping should be done from both ends in toward the middle, so there is no danger of tearout. The mortise ends can be only slightly undercut. The tenon must be carefully fitted since it will be visible on one face. Checking the fit of the tenon from both faces of the mortise is one way of working toward a good fit.

Rounding a tenon

As with most things, rounding the tenon to match the ends of the mortise gets easier after you do a few dozen of them. My first efforts were pretty slow as I would check each tenon end carefully, removing only small amounts of material before the next check. Now I grab my bastard file and a chisel and get to it.

One useful trick is to mark the centerline of the tenon in its thickness to help you guide your shaping (see the photo above). A steady hand, your discerning eye, and a pencil are all that's required. Forget about measuring to center. Start doing this by eye. You'll find that you get better at it with practice. And if you're a little off here it won't matter.

Put the board in a vise, and use a bastard file to begin knocking off the corners of the tenon. A file pass near the shoulder of the tenon will prevent any errant strokes of the file from nicking the tenon shoulder. Round the tenon up to this shoulder cut and finish off the rounding with a chisel. You can make a template for checking your progress by cutting a mortise into a piece of thin scrap, then cutting that in half (see the drawing at right).

I imagine the patient reader was waiting for the router trick to do this work. The tenon can be *partially* rounded over using a bearing-mounted roundover bit with a radius that's half the thickness of the tenon. So for a ½-in. wide tenon, set up a ¼-in. roundover bit in the router table. Take a pass just up to the tenon shoulder (see the photo at right). You have to be aware of the full diameter of the bit as it's spinning and stop just before the shoulder for two of the passes. The other two corners of the tenon are rounded by starting with the bit closer to the shoulder.

Another method would be to mount a fence with a recess that covers up the bit. Add a stop to the fence to prevent the bit from cutting too far. Decide for yourself about the trade-offs in time and anxiety. All of these router cuts will only get you part of the way home, unfortunately. The rest of the tenon will need rounding as before, with a chisel.

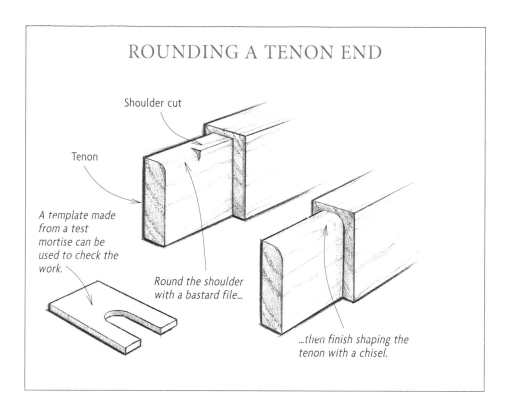

ROUNDING A TENON END

Shoulder cut

Tenon

A template made from a test mortise can be used to check the work.

Round the shoulder with a bastard file...

...then finish shaping the tenon with a chisel.

A roundover bit will get the tenon rounding started, but you have to be really careful to keep the router bit away from the tenon shoulder.

HAUNCHED TENONS

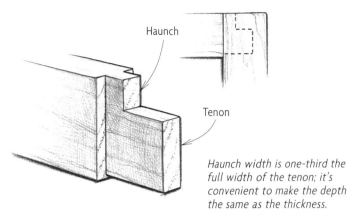

Haunch

Tenon

Haunch width is one-third the full width of the tenon; it's convenient to make the depth the same as the thickness.

STRAIGHT HAUNCH

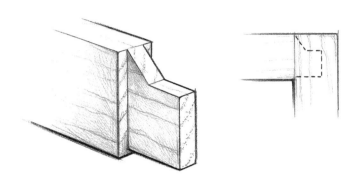

SECRET, OR MITERED, HAUNCH

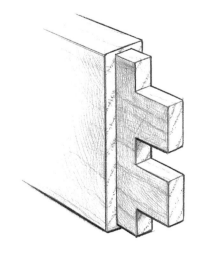

DOUBLE TENON WITH MULTIPLE HAUNCHES

HAUNCHED TENONS

Haunches provide strength and prevent twist in a rail without removing too much extra material from a mortise. There are some variations to the ways these haunches can be laid out (see the drawing at left), and the methods for cutting them depend on which one you make.

Straight haunches are simple right-angle cuts. I keep their depth the same size as their thickness for ease in laying out. Their width is about a third of the full width of the tenon, but this measurement can be adjusted to suit the requirements of the design.

With the different depth-stop turrets on a plunge router, the mortise part of the haunch is easily cut. The full mortise is set with one stop, and the haunch cut set on another. Take practice cuts to check these settings. The haunched tenon can be cut using either a backsaw, a bandsaw, or a table saw.

Secret, or mitered, haunches are cut on an angle. For this haunch even less material is removed from the mortised piece, and the cut is quick to make by hand with a chisel. The tenon haunch can be cut with a backsaw or on the bandsaw and cleaned up with a chisel.

Multiple haunches in wider stock are mortised in with the router. The tenon haunches are roughed in with the bandsaw or table saw. The end grain of the haunch can be cut on the router table with the board held flat. Stops prevent cutting into the double tenons.

MITERED AND NOTCHED TENONS

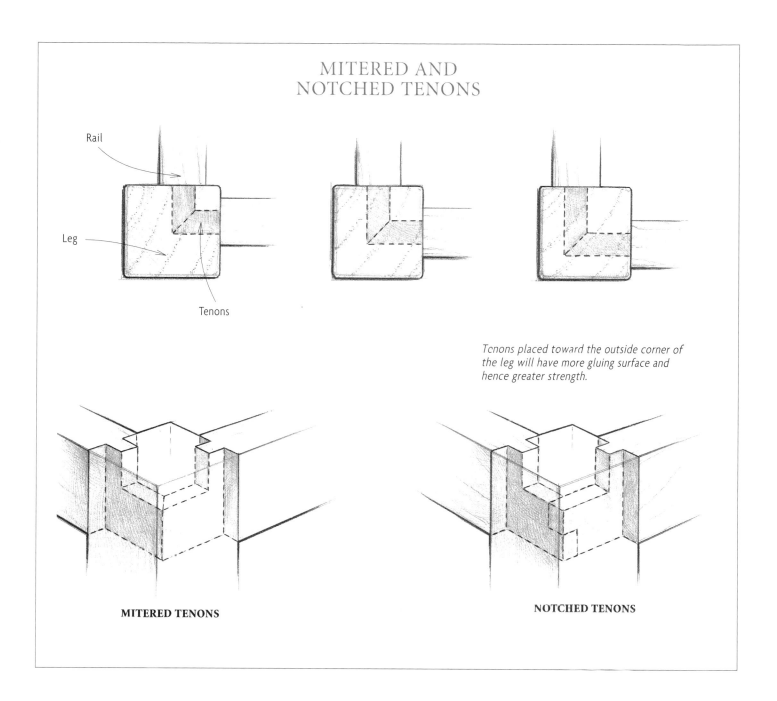

Rail

Leg

Tenons

Tenons placed toward the outside corner of the leg will have more gluing surface and hence greater strength.

MITERED TENONS

NOTCHED TENONS

MITERED AND NOTCHED TENONS

In many chair and table designs, the legs must provide room for tenons entering them from two different directions at the same level. To accommodate these tenons, mortises are cut to meet up inside a leg. As shown in the drawing above, the closer these mortises are to the outside of the leg, the deeper they can each become. And the deeper the mortise, the greater the gluing surface and the stronger the joint. The tenons can be mitered at their ends so they will fit to this depth, or they can be notched one over the other. In neither instance do I concern myself with getting these mating surfaces perfectly close to one another. They actually need some space between them to allow room for glue.

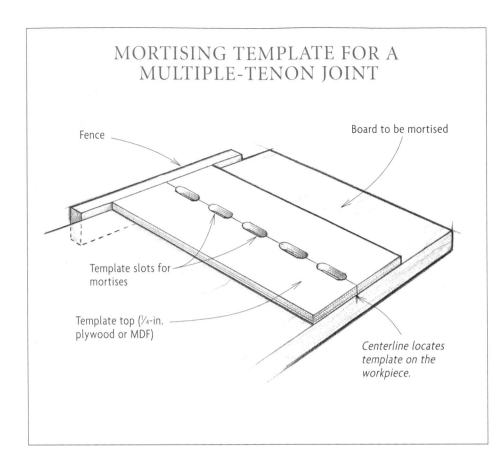

MORTISING TEMPLATE FOR A MULTIPLE-TENON JOINT

Fence

Board to be mortised

Template slots for mortises

Template top (1/4-in. plywood or MDF)

Centerline locates template on the workpiece.

For the one shown in the drawing at left, I used tenons spaced identically from one center tenon so the setup times were cut in half.

Cutting the tenon part of this joint can be done in two ways. You can use the router table with a crosscut jig, or you can make a tenon template that fits exactly into the mortises already cut. The tenon template can be cut on the router table or bandsaw or table saw. When it's made, mark out the tenons on the workpiece and rough-cut them on the bandsaw a slight 1/8 in. oversize. Then stick the template onto the workpiece with double-stick tape, and use a flush-trimming bit with a hand-held router to trim the tenons to size. The rounded corners are easily cleaned up with a chisel.

LOOSE TENONS

Loose tenons

Loose tenons fit into mortises cut into both pieces. Make the tenons a little shorter than the combined length of the two mortises.

MULTIPLE TENONS

When wide panels are used, the strength of multiple mortise-and-tenon joints makes them a good option. The alternative of using one long dado joint offers poor gluing surfaces (long grain to end grain); multiple tenons maximize long-grain gluing surfaces.

A simple way of cutting mortises for multiple tenons is to make a template that can be placed over the panel to be cut. It should have a fence on it that will locate the template at right angles to the board. Holes for a template guide are cut into it and should be accurately spaced for the mortises. Making the template is really the toughest part of the making the joint. It will take time to set up and cut correctly on the router table.

LOOSE TENONS

Situations will arise in the shop where an integral mortise and tenon will just be too difficult to cut. For instance, long rails on the headboard of a bed or angled tenons in a frame can be more easily mortised than tenoned. So the easy solution is to mortise both workpieces and put what's called a loose tenon between them (see the drawing at left). The mortises are cut using the same jig or template, and loose tenon stock is milled to fit these cuts.

Plunge-router mortising

One method of producing these cuts uses a plunge router with a straight fence. The router can cut mortises to depth quickly and accurately. Stops should be pencil-marked out on the board being cut, or actual stops can be clamped right onto the piece. A mortising jig can also be used to make these mortises. There's a problem, however, with many of these setups: When routing into the end grain of a workpiece, there is far less surface area to locate the router base against.

A simple solution is to create templates for these mortises (see pp. 155-159). Once made they can be used in a variety of situations. Their advantage for loose-tenon work is that they can be placed over long grain or end grain and provide a good working platform for the router base (see the photo at right). The mortise will be marked onto the piece being cut to reference the template. The template should be placed on matching face sides of each board so that both mortises line up properly. If their face sides are offset, a shim can be placed between the template fence and the narrower board to make up the difference.

A template clamped onto the end of a long rail locates the mortises for loose tenons and also provides a wide enough surface for the router to sit on.

Preparing the tenon stock

I make up lengths of stock to be used as loose tenons. These are easier and safer to handle than small individual pieces. I keep the tenon width just under the width of the mortise to make the fitting simple, but I make the tenon stock a hair thick to allow for any tweaking that may be required. Mortises routed into the end grain tend to be slightly larger than mortises routed into the long grain of the exact same board. That's because in end grain the bit tends to wander a little as it plunges to depth. If the tenon-stock thickness is a little oversize, there's room to fit these tenons.

The tenon stock should be rounded over before cutting to length. A roundover bit with a radius half the thickness of the tenon stock will do the job. Set the bit into the router table with the height set to cut all the edges of the tenon, producing matching rounded sides to the tenon. Scoring cuts can be made into the tenon stock so glue can escape when assembling. This is easily done on the table saw or with a straight bit set into the router table before crosscutting the tenons to final length.

I prefer to keep life simple in the woodshop, so I always glue in the loose tenon to the rail first. I usually bring it home with a dead-blow hammer or a clamp, then check its length. A tenon that needs trimming gets cut to length before it's glued into the mating piece.

A furniture project may begin as a gleam in the maker's eye, or a passing thought before going to sleep. As it evolves through sketches and plans and eventually to a piece of furniture, joinery plays a critical role in terms of design and execution. Here's an example of the process, using leg-and-apron construction to build the simple hall table shown in the photo below.

Certain joinery decisions must be made about a design even before beginning to mill up stock. For example, I can make a four-legged end table using stub tenons and slot dovetails for joints if panels will be used between the top and bottom rails. But if I want a simpler look to a piece, I prefer aprons with mortise-and-tenon joints (see the drawing at right on the facing page). The square table legs won't have a lot of room inside of them so after drawing out the table design, I draw out the joinery full scale to see what I've got to work with.

Accurate millwork is crucial to joinery success. I prefer to rough-mill 1/8 in. oversize in thickness and width, stickering my boards and letting the wood move if it wants to. Then I mill my boards straight and true to their final dimensions. I lay out all the legs and rails to arrange them for the best look and to number all the joints. Flame patterns in flatsawn lumber should complement the flow of the piece and any shaping work. Legs are arranged with their quartersawn faces matching.

ROUTING THE MORTISES

I start my mortising work by laying out one joint on a leg and clamping it into the mortising jig. I make sure the clamps don't get in the way of the router fence. Spacer blocks help raise the leg higher in the jig. I also clamp an end stop to locate the rest of the mortises.

Because the mortises meet up inside the joint, the bit depth cannot be set too deep, or it will cut into the side wall of the second mortise. Set for a slightly shallow cut, the bit leaves a little step at the inside corner of the mortise that is easily cleaned out with a chisel (see the drawing at left on the facing page).

I set the router for the haunch depth on a second depth stop, and also set up stops for the length of the full mortise and haunch.

I clamp on one stop for the end of the full mortise cut and another for the top of the haunch cut. I just pencil-mark the other stop for the full mortise. To begin routing, the depth stop is set for the haunch depth. Passes are made for the full length of the mortise, and then the depth setting is changed. To maximize the gluing surface for the tenons, I chop these mortises square and get just a bit more surface to glue that way.

CUTTING THE TENONS

I'll cut the tenons held flat on my router table. Ganging up two rails together gives better support against the fence. I take

practice cuts to check them. Even with all my cautious setups, I cut the tenons with light downward pressure. Pushing down hard on them will yield a minutely deeper cut, which may be all that's required to make these tenons too loose.

On the other hand, if they just miss fitting, then another pass with that downward push on them might be just enough to make them fit. If my luck has been bad all day, I'll use paper shims as well to sneak up on that good fit. My bullnose plane isn't too far away either for taking a slight shaving or two for a final fit.

The haunches are cut on the bandsaw with a fence to index all the cuts. After they're cut, I can fit the tenons all the way into their separate mortises. A tenon that has trouble fitting the mortise may have a haunch that won't fit. To check this, I simply flip the rail around and place the haunch in its mortise slot. The fit of the haunch then is all that is being checked.

If a tenon and its shoulders are not seating properly against the table leg, check the length of the tenon and the length of the haunch. Sometimes the haunch is not cut back far enough to allow the tenon full entry into the mortise. When each tenon does fit, then I trim their ends at a 45° angle on the table saw.

Hall table in alder.

ASSEMBLING THE TABLE

Perhaps the most overlooked job in the shop is dry assembly. By the time a project gets to the point of gluing up, a woodworker's patience has often grown wings and gone south. But taking the time to dry-assemble will pay off in several ways. First the order in which the piece goes together best can be seen. Where do the clamps fit, are extra clamps needed, which piece goes where and when? You can work out the trouble spots as you go rather than with the glue setting up and time running out.

I glue up the table in halves, two legs and a long rail at a time, which makes for an easy assembly. Laying a straightedge across the leg faces tells me if my clamping pressure is correct. Often, the placement of a clamp can pull an assembly out of true. This will make the next mortises harder to line up properly. Use a judicious amount of glue, and clean up

any excess before it stains the walls of the unglued mortise. If you have the assembly arranged with the unglued mortise facing up, the glue won't run into it.

When the glue has set up in these two assemblies, I then clean up the tops of the rails to the leg tops. I always keep the long grain of the rails a little bit proud of the tops of the legs. That way I can set the assembly in the vise and hand plane the rail down to size. If a leg top ends up high, it's a lot more trouble to bring it and all the other legs down in height.

The final assembly is made after another dry run has occurred, just to be certain I have all my clamps and clamping blocks ready. When the glue has set on this assembly, I clean up my rails to the leg tops again. To attach a top, biscuit-joiner slots can be cut into the inside of the rails. The top is attached using metal or wooden tabletop fasteners.

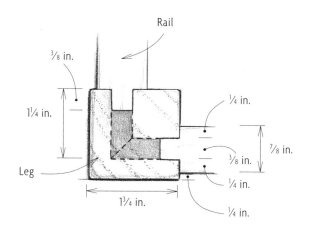

PLAN VIEW

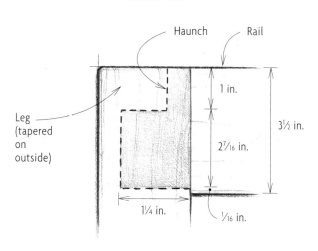

SIDE ELEVATION

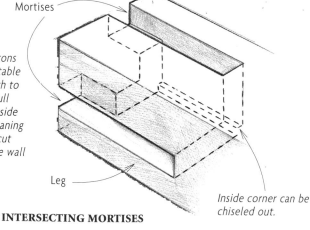

Mortises for the aprons meet up inside the table leg. Set the bit depth to cut just under the full depth of cut. The inside corner will need cleaning up, but too deep a cut would enter the side wall of the intersecting mortise.

INTERSECTING MORTISES

Inside corner can be chiseled out.

CUTTING LIST

Legs (4)	1¾ x 1¾ x 32
Long rails (2)	⅞ x 3½ x 42½
Side rails (2)	⅞ x 3½ x 18½
Top	¾ x 23½ x 51
Beading strips (2)	³⁄₁₆ x 1 x 40
Beading strips (2)	³⁄₁₆ x 1 x 16

(Dimensions are in inches.)

INDEX

Publisher: JAMES P. CHIAVELLI

Acquisitions Editor: RICK PETERS

Publishing Coordinator: JOANNE RENNA

Editor: RUTH DOBSEVAGE

Designer/Layout Artist: HENRY ROTH

Photographer: GARY ROGOWSKI, except where noted

Illustrator: BOB LA POINTE

Indexer: HARRIET HODGES

Typeface: AGENDA

Paper: WARREN PATINA MATTE, 70 lb., NEUTRAL pH

Printer: QUEBECOR PRINTING/HAWKINS, CHURCH HILL, TENNESSEE